AF711494

SciQUEST

Milo's Mission

SciQuest: Milo's Mission

First Edition 2026
Copyright @ Elemental Science, Inc.
Email: support@elementalscience.com

ISBN # 978-1-953490-35-3

Printed In USA For worldwide distribution

For more copies write to :
Elemental Science
PO Box 4583
Maryville, TN 37802
support@elementalscience.com

Copyright Policy

Quick Start Guide

It's time to head out on a quest—a scientific quest—a SciQuest.

What Will We Learn?

This semester, you will follow Milo as he makes friends in River Falls and learn about cells, vertebrates, planet Earth, and more. See p. 13 for a list of the topics explored in this program.

What Do I Need?

In addition to this activity guide, you will need the following materials:

1. **Novel –** *The PB Nanner Fluffer Test* - Each week, a chapter in this book will spark the student's interest to learn more. You can get the paperback novel, the Kindle version, or the audiobook.
2. **Age-appropriate Encyclopedias –** The bulk of the scientific information will come from these. We recommend that you have at least one or more of these on hand. If you can only purchase one or two of these, we recommend getting the ones with an *.
    - *Usborne Science Encyclopedia**
    - *Kingfisher Science Encyclopedia**
    - *Usborne Illustrated Dictionary of Science (used in Unit 1 and 2)*
    - *Super Earth Encyclopedia (used in Unit 3)*
    - *Everything You Need to Ace Science in One Big Fat Notebook*
3. **Student Materials –** You will need to purchase *SciQuest: Level 100 - Milo's Mission.* See inside the lab journal on p. 8 of this guide.
4. **Demonstration Supplies –** See p. 14 for a full list, or save yourself time and get the *SciQuest Year 1 Experiment Kit*, which includes the materials for both level 100 and 200.

What Will a Week Look Like?

Each week, you and your middle school student will:

- ✓ **Read** a chapter in a middle-grade fiction book with science hints and pick one or more of the assigned pages, discussing the information with the included questions.
- ✓ **Do** the weekly experiments by following the procedure, recording your observations and results, and drawing conclusions about what you saw.
- ✓ **Write** down what the students have learned in a way that is appropriate for their skills and label a related sketch.

You can also add in the memory work and additional science activities if you want to dig deeper into a topic. For a more detailed explanation of the components in each lesson, we highly recommend checking out the peek inside this program on pp. 6-7 and reading the introduction starting on pp. 9-12. Otherwise, the first lesson begins on p. 15.

Table of Contents

Teacher Notes At-A-Glance

The teacher guide is your go-to resource for guiding your students through science!

1. Two Scheduling Options

Provide guidance to the students with these two grid-style scheduling options. There are two-day-a-week and five-day-a-week schedules. These schedules break down the essential work and the optional activities into manageable chunks so that you can help them proceed with confidence.

2. Notes

Get key information and further explanations of the topics when necessary after the schedules. These notes will assist you as you teach science to your students and are described below.

Read Science

3. Reading Assignments

See the week's reading assignments. First, you will find the chapter from the springboard novel. After that, you will find the encyclopedia pages, which contain the core of the information for the week.

4. Discussion Questions with Answers

Get discussion questions with the answers so you have the tools necessary to lead an effective discussion time without having to read the selection.

5. Optional Readings

Find optional books you can check out from your local library.

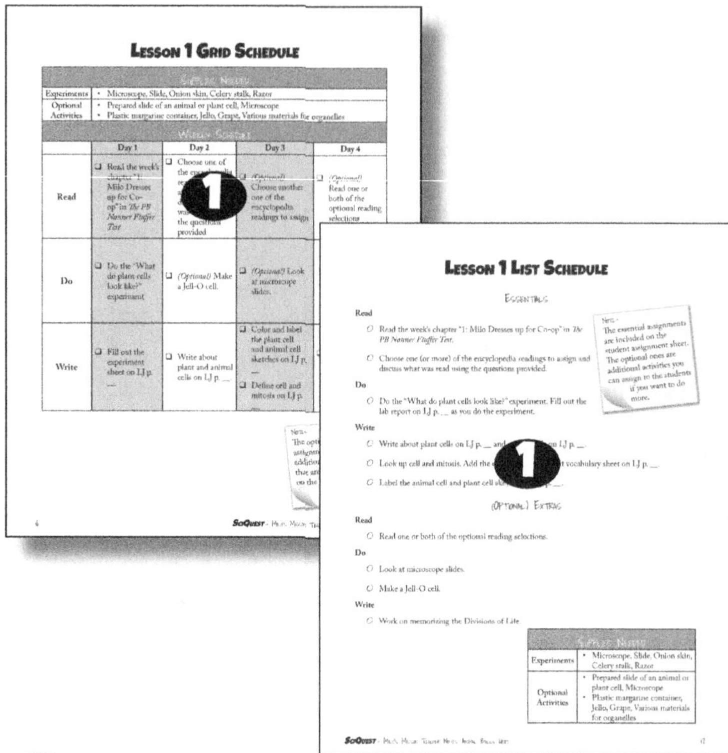

Lesson 1 Grid Schedule

Lesson 1 List Schedule

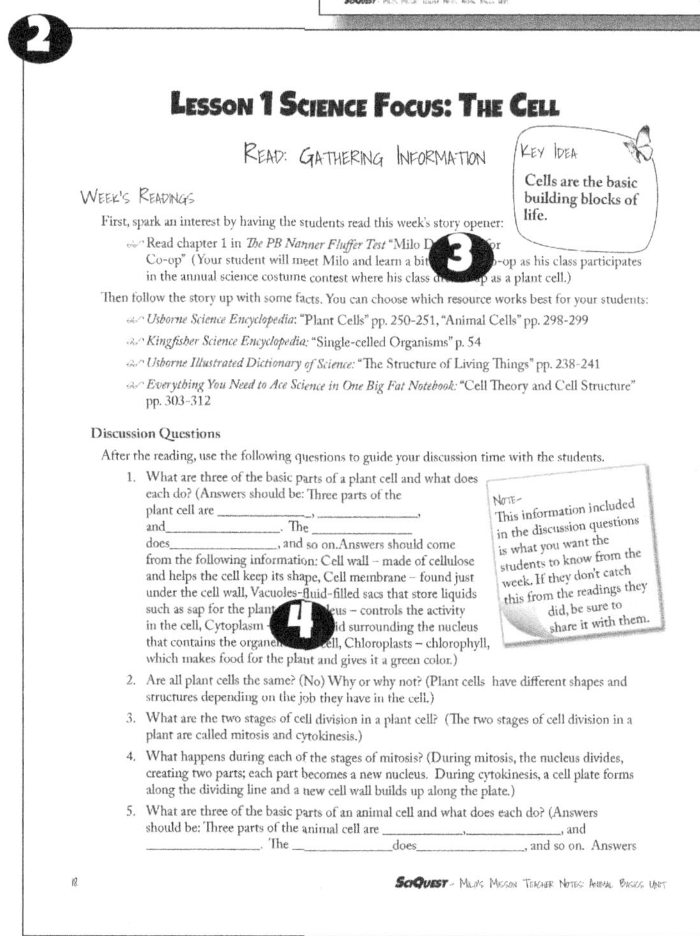

Lesson 1 Science Focus: The Cell

Read: Gathering Information

Key Idea: Cells are the basic building blocks of life.

Week's Readings

First, spark an interest by having the students read this week's story opener:

Read chapter 1 in *The PB Nanner Fluffer Test* "Milo D… for Co-op" (Your student will meet Milo and learn a bit … -op as his class participates in the annual science costume contest where his class … as a plant cell.)

Then follow the story up with some facts. You can choose which resource works best for your students:

- *Usborne Science Encyclopedia*: "Plant Cells" pp. 250-251, "Animal Cells" pp. 298-299
- *Kingfisher Science Encyclopedia*: "Single-celled Organisms" p. 54
- *Usborne Illustrated Dictionary of Science*: "The Structure of Living Things" pp. 238-241
- *Everything You Need to Ace Science in One Big Fat Notebook*: "Cell Theory and Cell Structure" pp. 303-312

Discussion Questions

After the reading, use the following questions to guide your discussion time with the students.

1. What are three of the basic parts of a plant cell and what does each do? (Answers should be: Three parts of the plant cell are ______, ______, and ______. The ______ does ______, and so on. Answers should come from the following information: Cell wall – made of cellulose and helps the cell keep its shape, Cell membrane – found just under the cell wall, Vacuoles–fluid-filled sacs that store liquids such as sap for the plant, …us – controls the activity in the cell, Cytoplasm … id surrounding the nucleus that contains the organe… cell, Chloroplasts – chlorophyll, which makes food for the plant and gives it a green color.)
2. Are all plant cells the same? (No) Why or why not? (Plant cells have different shapes and structures depending on the job they have in the cell.)
3. What are the two stages of cell division in a plant cell? (The two stages of cell division in a plant are called mitosis and cytokinesis.)
4. What happens during each of the stages of mitosis? (During mitosis, the nucleus divides, creating two parts; each part becomes a new nucleus. During cytokinesis, a cell plate forms along the dividing line and a new cell wall builds up along the plate.)
5. What are three of the basic parts of an animal cell and what does each do? (Answers should be: Three parts of the animal cell are ______, ______, and ______. The ______ does ______, and so on. Answers

Note – This information included in the discussion questions is what you want the students to know from the week. If they don't catch this from the readings they did, be sure to share it with them.

Teacher Notes At-A-Glance

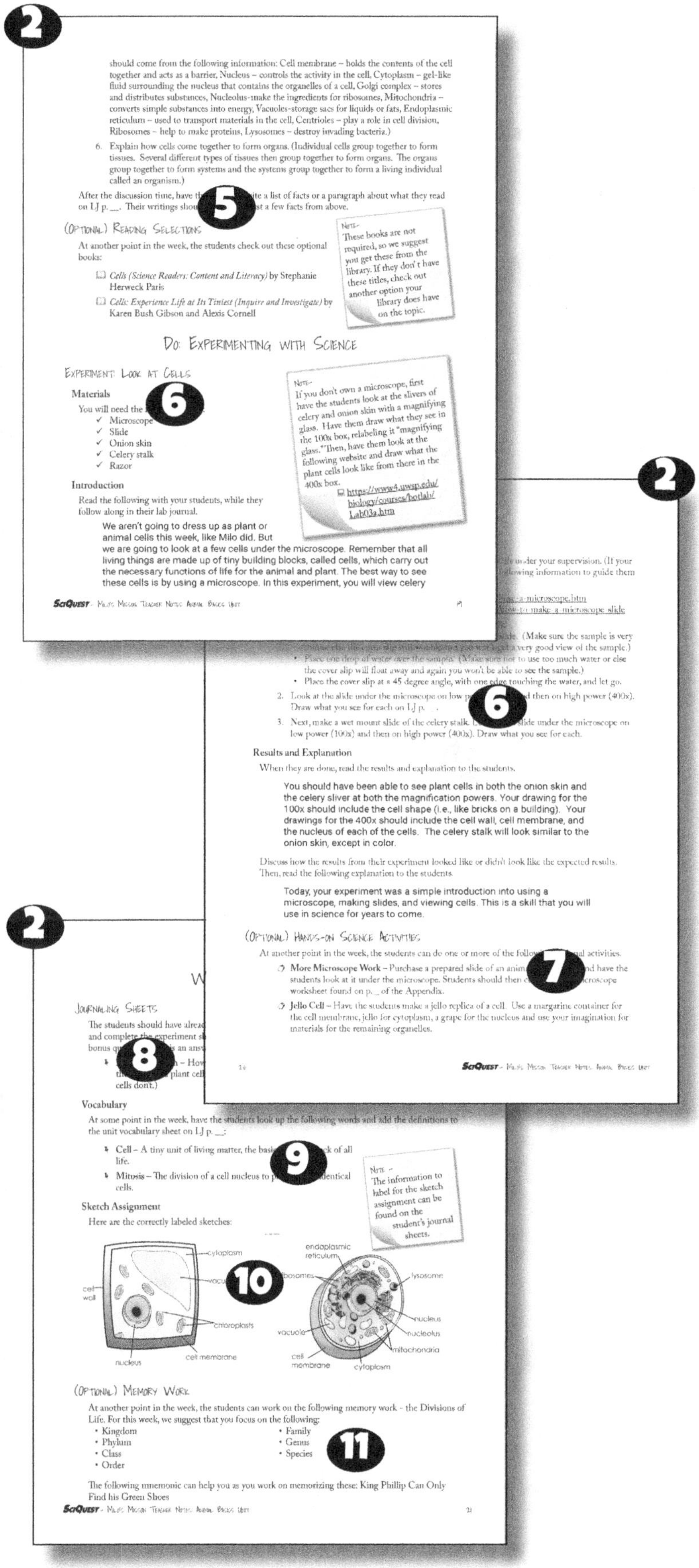

Do Science

6. Experiment Information

Know what you need to know about the weekly experiment, which will coordinate with the topic. This section includes the introduction and procedure, which are also included in the student materials. It also lists the expected results, an explanation, and ideas to take the hands-on science fun further.

7. Optional Hands-on Activities

Add a bit of fun with these optional research projects and hands-on science project ideas. These relate to and reinforce what the students are learning about the week's subject.

Write Science

8. Journal Sheets

Understand what the students will be writing down for the week, along with a bonus question. These are deeper inquiries designed to make the students think about the subject matter.

9. Vocabulary

Enhance your students' science vocabulary by introducing words that are relevant to the weekly topic.

10. Sketch Answers

Know how and where the students should label the pre-drawn sketches with the completed images.

11. Optional Memory Work

Have the students memorize key facts about the unit's topics with these optional memory work selections.

Lab Journal At-A-Glance

Go on a scientific quest to learn about cells, animals, and our planet.

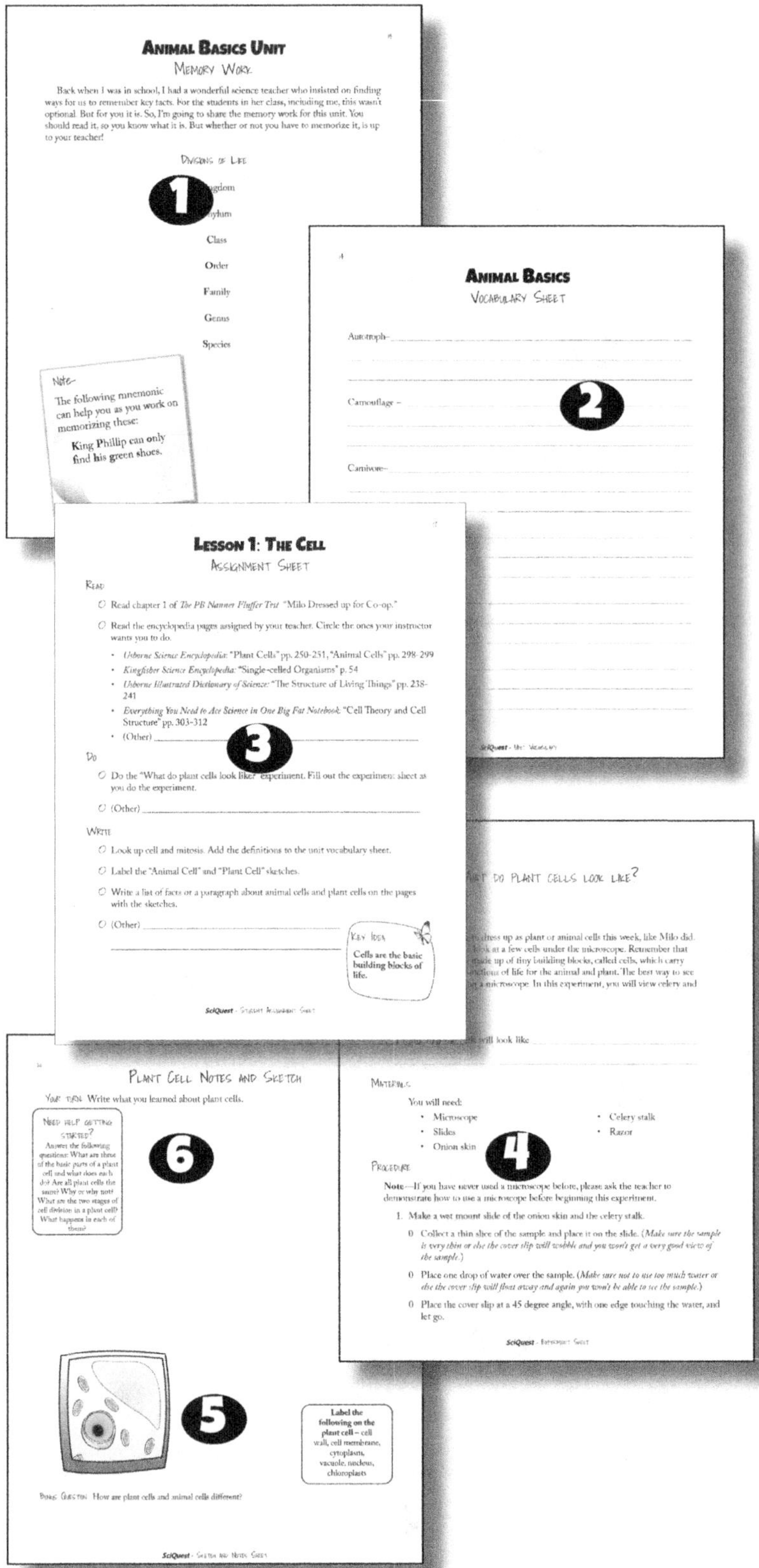

1. Unit Memory Work

Memorize key facts about the unit's topics with these optional memory work selections.

2. Unit Glossary

Keep all the unit vocabulary in one place. The glossary gives space to write the definitions for each vocabulary word.

3. Student Assignment Sheet

See the plan for the week on the student assignment sheets, broken into reading, doing, and writing sections. This sheet has all the tasks for the week with checkboxes for you to use as you complete the work.

4. Experiment Sheet

Do an experiment related to the topic for the week. The Experiment Sheet contains the introduction and the procedure for the experiment. It also has space to record the hypothesis, materials, observations, results, and conclusion.

5. Pre-drawn Sketch

Color the pre-drawn sketch relating to the week's topic. Then, label it with the information given on the page.

6. Written Assignment

Know the questions for discussion time, and have plenty of space to write either a list of facts, an outline, or a summary.

SciQuest: Milo's Mission
Introduction

Welcome to a quest to learn about science, or should I say a SciQuest! This is a journey that your students are going to embark on with a bit of independence.

In *SciQuest 100: Milo's Mission*, you'll be learning about basics blocks of life, vertebrates, and our planet. Milo Pemberton will serve as an introduction for your students for the week. They will learn about the topics in biology and earth science by reading, doing, and writing about science.

The Broad-Brush-Stroke View

In a nutshell, your students will enjoy learning about science as they do three things each week. These three things are based on the *Three Keys to Teaching Science* by Paige Hudson. Check them out here: https://elementalscience.com/blogs/news/3-keys. Here is what they are:

Read: Gathering Information

In this section of the lesson notes, you will find the assigned readings and discussion questions. Check out this podcast to understand the importance of discussion time:

💻 Don't skip that discussion time - https://elementalscience.com/blogs/podcast/53

You'll also find a few optional assignments for those students who want to dig deeper.

Do: Experimenting with Science

In this section of the lesson notes, we will be employing those inquiry-based methods. In other words, the students are going to do science. You will find an experiment introduction and directions – don't worry, these are in the student's journal, too. You will also find the expected results, an explanation, and any troubleshooting tips for the experiment in this section. Read this article to understand how your role is shifting from presenter-of-hands-on-science to mentor-of-a-budding-scientist:

💻 Scientific demonstrations vs. experiments - https://elementalscience.com/blogs/podcast/94

Plus, we added a few optional activities for those students who love to play with science.

Write: Keeping a Journal

In this section of the lesson notes, we are helping our students to organize what they have learned and to think analytically about it. This is where you will find the written assignments, vocabulary, and memory work for the lesson. The written assignments include a customizable writing assignment (either a list of facts, an outline, or a paragraph) and the sketch for the lesson. Watch this series of videos to learn about the progression of writing skills as they pertain to science:

💻 Writing in Homeschool Science: The Elementary Years - https://youtu.be/BrunFyeHhlQ
💻 Writing in Homeschool Science: Middle School and Beyond - https://youtu.be/B96DKXriCng

Also, with the written work, we have included a bonus question to help the students start to make those all-important connections.

The Nitty-Gritty Details of the Lesson Notes

Week's Readings

First, spark an interest by having the students read this week's story opener. These chapters come from the middle-grade novel, *The PB Nanner Fluffer Test*. Your students will read about Milo as he makes friends at his local homeschooling co-op. Each chapter has a hint of science and is designed to be read by the students on their own.

Later in the week, the students will follow up the story with facts. The following encyclopedias have pages detailed in this guide. You can choose which resources work best for your student. They will be used in subsequent levels of this program. If you can only purchase one or two of these, we recommend getting the ones with an *.

- *Usborne Science Encyclopedia**
- *Kingfisher Science Encyclopedia**
- *Usborne Illustrated Dictionary of Science (used in Units 1 and 2)*
- *Super Earth Encyclopedia (used in Unit 3)*
- *Everything You Need to Ace Science in One Big Fat Notebook*

Discussion Questions

After all that reading, we have provided questions, along with the answers, to guide your discussion time with the students.

(Optional) Reading Selections

In this section, you will find additional reading suggestions. These are not required, but these library books can be used to enhance what the students have learned.

Experiment Information

Materials

This section will detail the supplies you need for the experiment. As much as possible, we have attempted to make these household supplies. However, if you don't want to pull these items together, you can purchase an experiment kit from us that contains the supplies for level 100 and 200.

Introduction

This section shares a glimpse of the science behind the experiment. It is also included in the lab journal.

The paragraphs written in this font are meant to be read to the student.

Directions

This section will detail what to do for the experiment, step by step. It is also included in the lab journal.

Results and Explanation

When the students are done with the experiment, read the results and explanation to them. Discuss how the results from their experiment looked like or didn't look like the expected results.

As before, the paragraphs written in this font are meant to be read to the student.

(Optional) Hands-on Science Activities

In this section, you will find optional hands-on science activities to reinforce what the students are learning.

Write: Keeping a Journal

Journaling Sheets

This section will have the page numbers and bonus questions. The students should record what they learn about the subjects on notes pages and the experiment sheets in the lab journal. On the notes page, the students can also answer the bonus question. These questions are meant to challenge the students and cause them to think about the subject.

Vocabulary

The students will look up the vocabulary words and add the definitions to their unit vocabulary sheet in the lab journal. In this guide, you will find the definitions for the words they have to look up.

Sketch

In this section, you will find the labeled sketches to use as an answer key.

(Optional) Memory Work

Every unit has memory work, which you can choose to assign or not.

Additional Resources

We have put together a page containing quick links to the activities suggested in this guide along with several helpful downloads:

💻 https://elementalscience.com/blogs/resources/sq100

How to Schedule this Study

SciQuest 100: Milo's Mission should take up to three hours per week to complete. You and your student can choose whether to complete the work over five days or over two days. Below are two options for scheduling to give you an idea of how you can schedule your week:

Two-days-a-week Schedule

Day 1

Define the vocabulary, do the experiment, and complete the experiment sheet.

Day 2

Read the assigned pages and discuss them together, take your notes, and complete the sketch.

Five-days-a-week Schedule

Day 1

Do the experiment and complete the experiment sheet.

Day 2

Record the dates and define the vocabulary.

Day 3

Read the assigned pages, discuss them, and complete the sketch.

Day 4

Prepare the science report or an outline.

Day 5

Complete one of the other activities assigned by your teacher and work on the memory work.

Final Thoughts

If you find that this program contains too much work, please tailor it to the needs of your students. With that said, I encourage you to contact us with any questions or problems that you might have concerning *SciQuest 100: Milo's Mission* at support@elementalscience.com. Our team will be more than happy to answer them as soon as we are able. I hope that you and your students enjoy your quest to learn about science!

~ Paige Hudson

List of Topics

Animal Basics Unit

- Plant Cells
- Animal Cells
- Microscopes
- Taxonomy
- Dichotomous Keys
- Migration
- Animal Navigation
- Adaptations
- Defenses
- Animal Diet (Herbivore, Carnivore, and Omnivore)

Vertebrates Unit

- Fish
- Amphibians
- Reptiles
- Birds
- Mammals
- Animal Reproduction
- Mitosis
- Meiosis

Scientist Biography Unit

- The students will learn about a scientist of their choice.

Planet Earth Unit

- Inside the Earth
- Maps and Mapping
- Longitude and Latitude
- Rivers
- Oceans
- Currents
- Glaciers
- Biomes
- Habitats
- Water Cycle
- Nitrogen Cycle
- Carbon Cycle

Demonstration Supplies

Animal Basics Unit

Lesson	Materials Needed
1	Microscope, Slide, Onion skin, Celery stalk, Razor
2	Leaf from outside, Dichotomous key for plants
3	A place to observe, Access to the Internet or a local field guide
4	Gloves, Tweezers, Wooden stick (coffee stirrer or popsicle stick will work), Newspaper or paper towel, Owl pellet
5	Newspaper, Plain paper, Black and green construction paper

Vertebrates Unit

Lesson	Materials Needed
6	1 Large clear glass jar or bowl, 3 Small balloons, 3 Small marbles, Ruler, Water
7	Chopsticks, Tweezers, Pliers, Eye dropper, Sugar, Water, Gummy worms, Peanuts, Seeds, Raisins, Plates
8	1 Frog dissection kit, 1 Preserved frog
9	Clothespin, Blindfold, 5 Plates, 5 Pieces of bread with different edible spreads on them (such as garlic, cinnamon sugar, butter, peanut butter, and plain)
10	Felt, Cotton balls, Water, 1 Large cup, 4 Small cups, Instant read thermometer

Planet Earth Unit

Lesson	Materials Needed
12	Permanent marker, 1 Raw egg, 1 Hard-boiled egg
13	Blue balloon (with the continents drawn or printed on it), Flat map, Pin
14	Pitcher for water, Water, Cookie sheet, Paper cup, Straw, Dirt or sand, Small rocks, Books, Tape
15	Aluminum bread pan or Plastic bin, Air dry clay, Water, Sand (1 cup), 2 Straws
16	Glacier Melt Model (You will need a cup, water, pebbles, and sand to make this), Large cutting board with a handle, Large rubber band
17	Paper towels, Cookie sheet, Wax paper, Rubber band, Water
18	Cup, Water, Plastic Baggie, Rubber Band

SciQuest

Milo's Mission Teacher Notes

Animal Basics Unit

What you'll study in this unit-

*LJ = Lab Journal

Lesson 1 Grid Schedule

Supplies Needed	
Experiments	• Microscope, Slide, Onion skin, Celery stalk, Razor
Optional Activities	• Prepared slide of an animal or plant cell, Microscope • Plastic margarine container, Jello, Grape, Various materials for organelles

Weekly Schedule

	Day 1	Day 2	Day 3	Day 4
Read	❑ Read Chapter 1: "Milo Dresses up for Co-op" in *The PB Nanner Fluffer Test*	❑ Choose one of the encyclopedia readings to assign and discuss what was read using the questions provided	❑ *(Optional)* Choose another one of the encyclopedia readings to assign	❑ *(Optional)* Read one or both of the optional reading selections
Do	❑ Do the "What do plant cells look like?" experiment	❑ *(Optional)* Make a Jell-O cell	❑ *(Optional)* Look at microscope slides	
Write	❑ Fill out the experiment sheet	❑ Write about plant and animal cells	❑ Color and label the plant cell and animal cell sketches ❑ Define cell and mitosis	❑ *(Optional)* Work on memorizing the Divisions of Life

Note-
The optional assignments are additional activities that are not included on the student's assignment sheet.

Lesson 1 List Schedule

Essentials

Read

- O Read Chapter 1: "Milo Dresses up for Co-op" in *The PB Nanner Fluffer Test*.
- O Choose one (or more) of the encyclopedia readings to assign and discuss what was read using the questions provided.

Note–
The essential assignments are included on the student assignment sheet. The optional ones are additional activities you can assign to the students if you want to do more.

Do

- O Do the "What do plant cells look like?" experiment. Fill out the lab report as you do the experiment.

Write

- O Write about plant cells and animal cells.
- O Look up cell and mitosis. Add the definitions to the unit vocabulary sheet.
- O Label the animal cell and plant cell sketches.

(Optional) Extras

Read

- O Read one or both of the optional reading selections.

Do

- O Look at microscope slides.
- O Make a Jell-O cell.

Write

- O Work on memorizing the Divisions of Life.

Supplies Needed	
Experiments	• Microscope, Slide, Onion skin, Celery stalk, Razor
Optional Activities	• Prepared slide of an animal or plant cell, Microscope • Plastic margarine container, Jello, Grape, Various materials for organelles

Lesson 1 Science Focus: The Cell

Read: Gathering Information

Key Idea

Cells are the basic building blocks of life.

Week's Readings

First, spark an interest by having the students read this week's story opener:

- Read Chapter 1 in *The PB Nanner Fluffer Test*: "Milo Dresses up for Co-op." (Your student will meet Milo and learn a bit about his co-op as his class participates in the annual science costume contest where his class dresses up as a plant cell.)

Then follow the story up with some facts. You can choose which resource works best for your students:

- *Usborne Science Encyclopedia*: "Plant Cells" pp. 250-251, "Animal Cells" pp. 298-299
- *Kingfisher Science Encyclopedia:* "Single-celled Organisms" p. 54
- *Usborne Illustrated Dictionary of Science:* "The Structure of Living Things" pp. 238-241
- *Everything You Need to Ace Science in One Big Fat Notebook:* "Cell Theory and Cell Structure" pp. 303-312

Discussion Questions

After the reading, use the following questions to guide your discussion time with the students.

1. What are three of the basic parts of a plant cell and what does each do? (Answers should be: Three parts of the plant cell are ______________, ________________, and_________________. The _______________ does________________, and so on. Answers should come from the following information: Cell wall – made of cellulose and helps the cell keep its shape, Cell membrane – found just under the cell wall, Vacuoles-fluid-filled sacs that store liquids such as sap for the plant cell, Nucleus – controls the activity in the cell, Cytoplasm – gel-like fluid surrounding the nucleus that contains the organelles of a cell, Chloroplasts – chlorophyll, which makes food for the plant and gives it a green color.)

> Note–
> This information included in the discussion questions is what you want the students to know from the week. If they don't catch this from the readings they did, be sure to share it with them.

2. Are all plant cells the same? (No.) Why or why not? (Plant cells have different shapes and structures depending on the job they have in the cell.)
3. What are the two stages of cell division in a plant cell? (The two stages of cell division in a plant are called mitosis and cytokinesis.)
4. What happens during each of the stages of mitosis? (During mitosis, the nucleus divides, creating two parts; each part becomes a new nucleus. During cytokinesis, a cell plate forms along the dividing line and a new cell wall builds up along the plate.)
5. What are three of the basic parts of an animal cell and what does each do? (Answers should be: Three parts of the animal cell are _____________,_______________, and _________________. The _______________does_________________, and so on. Answers

should come from the following information: Cell membrane – holds the contents of the cell together and acts as a barrier, Nucleus – controls the activity in the cell, Cytoplasm – gel-like fluid surrounding the nucleus that contains the organelles of a cell, Golgi complex – stores and distributes substances, Nucleolus – make the ingredients for ribosomes, Mitochondria – converts simple substances into energy, Vacuoles – storage sacs for liquids or fats, Endoplasmic reticulum – used to transport materials in the cell, Centrioles – play a role in cell division, Ribosomes – help to make proteins, Lysosomes – destroy invading bacteria.)

6. Explain how cells come together to form organs. (Individual cells group together to form tissues. Several different types of tissues then group together to form organs. The organs group together to form systems and the systems group together to form a living individual called an organism.)

After the discussion time, have the students write a list of facts or a paragraph about what they read. Their writings should include at least a few facts from above.

(Optional) Reading Selections

At another point in the week, the students can check out these optional books:

- *Cells (Science Readers: Content and Literacy)* by Stephanie Herweck Paris
- *Cells: Experience Life at Its Tiniest (Inquire and Investigate)* by Karen Bush Gibson and Alexis Cornell

Note–
These books are not required, so we suggest you get these from the library. If they don't have these titles, check out another option your library does have on the topic.

Do: Experimenting with Science

Experiment: Look at Cells

Materials

You will need the following items:

- ✓ Microscope
- ✓ Slide
- ✓ Onion skin
- ✓ Celery stalk
- ✓ Razor

Note–
If you don't own a microscope, first have the students look at the slivers of celery and onion skin with a magnifying glass. Have them draw what they see in the 100x box, relabeling it "magnifying glass." Then, have them look at the following website and draw what the plant cells look like from there in the 400x box.

https://www4.uwsp.edu/biology/courses/botlab/Lab03a.htm

Introduction

Read the following with your students, while they follow along in their lab journal.

> We aren't going to dress up as plant or animal cells this week, like Milo did. But we are going to look at a few cells under the microscope. Remember that all living things are made up of tiny building blocks, called cells, which carry out the necessary functions of life for the animal and plant. The best way to see these cells is by using a microscope. In this experiment, you will view celery

and onion cells.

Directions

Have the students do the following experiment looking at plant cells under your supervision. (If your students have never used a microscope before, please refer to the following information to guide them through using a microscope.)

- How to Use a Microscope: https://www.youtube.com/watch?v=Kp28CsflBp8
- https://elementalscience.com/blogs/science-activities/how-to-make-a-microscope-slide

1. Make a wet mount slide of the onion skin and celery stalk.
    - Collect a thin slice of the sample and place it on the slide. (Make sure the sample is very thin or else the cover slip will wobble and you won't get a very good view of the sample.)
    - Place one drop of water over the sample. (Make sure not to use too much water or else the cover slip will float away and again you won't be able to see the sample.)
    - Place the cover slip at a 45 degree angle, with one edge touching the water, and let go.
2. Look at the slide under the microscope on low power (100x) and then on high power (400x). Draw what you see for each.
3. Next, make a wet mount slide of the celery stalk. Look at the slide under the microscope on low power (100x) and then on high power (400x). Draw what you see for each.

Results and Explanation

When they are done, read the results and explanation to the students.

> You should have been able to see plant cells in both the onion skin and the celery sliver at both the magnification powers. Your drawing for the 100x should include the cell shape (i.e., like bricks on a building). Your drawings for the 400x should include the cell wall, cell membrane, and the nucleus of each of the cells. The celery stalk will look similar to the onion skin, except in color.

Discuss how the results from their experiment looked like or didn't look like the expected results. Then, read the following explanation to the students.

> Today, your experiment was a simple introduction into using a microscope, making slides, and viewing cells. This is a skill that you will use in science for years to come.

(Optional) Hands-on Science Activities

At another point in the week, the students can do one or more of the following optional activities.

- **More Microscope Work** – Purchase a prepared slide of an animal or plant cell and have the students look at it under the microscope. Students should then complete the microscope worksheet found on p. 135 of the Appendix.
- **Jello Cell** – Have the students make a jello replica of a cell. Use a margarine container for the cell membrane, jello for cytoplasm, a grape for the nucleus and use your imagination for materials for the remaining organelles.

Write: Keeping a Journal

Journaling Sheets

The students should have already recorded what they learned about cells on their notes page (LJ pp. 20-21) and completed the experiment sheet (LJ pp. 18-19). On the notes page, the students can also answer the bonus question. Here is an answer for your reference as you discuss this with your students.

- **Bonus Question** – How are plant and animal cells different? (Plant cells have cell walls and the inside of a plant cell is under pressure. Plant cells also contain chloroplasts, while animal cells don't.)

Vocabulary

At some point in the week, have the students look up the following words and add the definitions to the unit vocabulary sheet on LJ pp. 14-16:

- **Cell** – A tiny unit of living matter, the basic building block of all life.
- **Mitosis** – The division of a cell nucleus to produce two identical cells.

Note - The information to label for the sketch assignment can be found on the student's journal sheets.

Sketch Assignment

Here are the correctly labeled sketches: (LJ pp. 20-21)

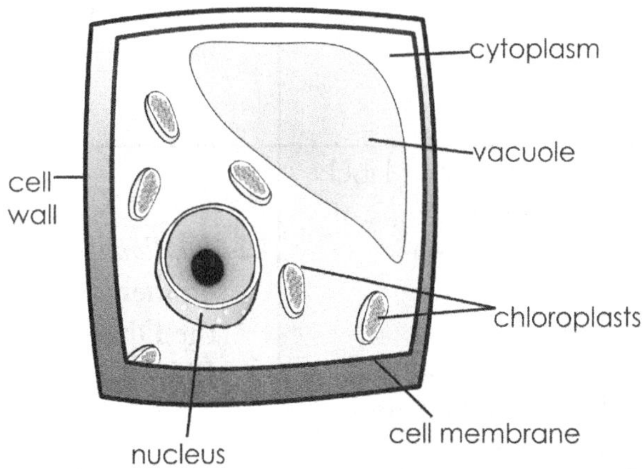

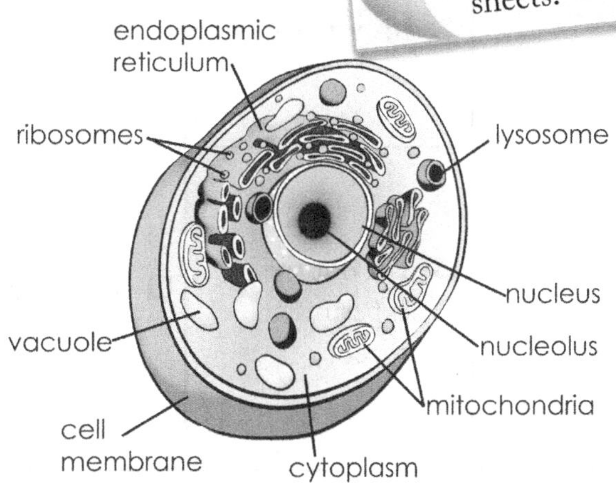

(Optional) Memory Work

At another point in the week, the students can work on the following memory work - the Divisions of Life. For this week, we suggest that you focus on the following:

- Kingdom
- Phylum
- Class
- Order
- Family
- Genus
- Species

The following mnemonic can help you as you work on memorizing these: King Phillip Can Only Find his Green Shoes.

Lesson 2 Grid Schedule

Supplies Needed	
Experiments	• Leaf from outside, Dichotomous key for plants
Optional Activities	• Another leaf, Field guide for trees • Paper, Pencil

Weekly Schedule				
	Day 1	**Day 2**	**Day 3**	**Day 4**
Read	❑ Read Chapter 2: "Milo Unpacked... With Help" in *The PB Nanner Fluffer Test*	❑ Choose one of the encyclopedia readings to assign and discuss what was read using the questions provided	❑ *(Optional)* Choose another one of the encyclopedia readings to assign	❑ *(Optional)* Choose another one of the encyclopedia readings to assign
Do	❑ Do the "What kind of tree is it?" experiment	❑ *(Optional)* Make a Dichotomous Key	❑ *(Optional)* Do the New Species activity	❑ *(Optional)* Create a Tree Classification book
Write	❑ Fill out the experiment sheet	❑ Write about classification	❑ Color and label the divisions of life sketch ❑ Define classification, kingdom, and species	❑ *(Optional)* Work on memorizing the Divisions of Life

Lesson 2 List Schedule

Essentials

Read

- ○ Read Chapter 2: "Milo Unpacked… With Help" in *The PB Nanner Fluffer Test.*
- ○ Choose one of the encyclopedia readings to assign and discuss what was read using the questions provided.

Do

- ○ Do the "What kind of tree is it?" experiment. Fill out the lab report as you do the experiment.

Write

- ○ Write about classification.
- ○ Look up classification, kingdom, and species. Add the definitions to the unit vocabulary sheet.
- ○ Label the divisions of life sketch.

(Optional) Extras

Read

- ○ Choose another one (or two) of the encyclopedia readings to assign.

Do

- ○ Create a Tree Classification book or make a Dichotomous Key.
- ○ Make a Dichotomous Key.
- ○ Do the New Species activity.

Write

- ○ Work on memorizing the Divisions of Life.

Supplies Needed	
Experiments	• Leaf from outside, Dichotomous key for plants
Optional Activities	• Another leaf, Field guide for trees • Paper, Pencil

Lesson 2 Science Focus: Classification

Read: Gathering Information

> **Key Idea**
>
> **All living things can be organized or grouped.**

Week's Readings

First, spark an interest by having the students read this week's story opener:

- Read Chapter 2 in *The PB Nanner Fluffer Test*: "Milo Unpacked... With Help." (Your student will watch Milo work with his mother to unpack and organize his moving boxes.)

Then follow the story up with some facts. You can choose which resource works best for your students:

- *Usborne Science Encyclopedia*: "Classifying Plants" pp. 294-295, "Classifying Animals" pp. 340-343
- *Kingfisher Science Encyclopedia*: "Classification of Living Things" pp. 52-53
- *Usborne Illustrated Dictionary of Science:* "The Classification of Living Things" pp. 340-341
- *Everything You Need to Ace Science in One Big Fat Notebook:* "Organisms and Biological Classification" pp. 292-300

Discussion Questions

After all that reading, use the following questions to guide your discussion time with the students.

1. What are the five main kingdoms? (The five kingdoms are monera, protista, fungi, plant, and animal.)
2. How do scientists classify living things? (Scientists classify living things by identifying their main characteristics and then comparing them with similar species.)
3. What are the two main divisions of the plant kingdom? (The two main divisions of the plant kingdom are plants that contain vascular tissue and plants that do not.)
4. What are the taxonomic ranks (or divisions of life)? (Kingdoms can be broken down into phyla, each phylum is broken down into classes, each class is broken down into orders, each order is broken down into families, each family is broken down into genera, and each genus is broken down into species.)
5. What language is a biological name given in and how is it created? (Biological names are always written in Latin. The first part of a biological name is based on the organism's genus, the second identifies its species.)

After the discussion time, have the students write a list of facts or a paragraph about what they read. Their writings should include at least a few facts from above.

(Optional) Reading Selections

At another point in the week, the students can check out these optional books:

- *Karl, Get Out of the Garden!: Carolus Linnaeus and the Naming of Everything* by Anita Sanchez and Catherine Stock

📖 *Animal Classification: Do Cats Have Family Trees? (Show Me Science)* by Eve Hartman

Do: Experimenting with Science

Experiment: What kind of tree is it?

> Note -
> If you don't want to type in the whole long dichotomous key link, we have got a solution for you! Check out the easy-to-click resource page for this program:
> 💻 https://elementalscience.com/blogs/resources/sq100

Materials

You will need the following items:

- ✓ Leaf from outside
- ✓ Dichotomous key for plants from the following website:
  💻 https://treespnw.forestry.oregonstate.edu/

Introduction

Read the following with your students, while they follow along in their lab journal.

> A dichotomous key is a method that can be used to identify a living thing. At each step, you are given a question that has only two answers. The answers eventually lead to the identity of the living thing. In this experiment, you will be using a dichotomous key from the internet to identify an unknown leaf.

Directions

Have the students do the following experiment to use a leaf to learn about a local tree under your supervision.

1. Go outside and choose a leaf from a tree that you want to identify.
2. Once inside, write down what kind of tree you think the leaf is from.
3. Spend a few moments observing the leaf, making notes about its shape and size under the "Observations" section.
4. Use the dichotomous key from the website above to identify the leaf. (Although the website is for trees of the Pacific Northwest, it will help you identify trees from all over the USA and other countries. If you don't live in America and the key doesn't work for your leaf, just head over to your search engine of choice and type in "Tree Dichotomous Key for {the place you live}.)

Results and Explanation

When they are done, read the results and explanation to the students.

> You should have been able to identify the genus of tree that the leaf came from.

Discuss how the results from their experiment looked like or didn't look like the expected results. Then, read the following explanation to the students.

> This was a simple introduction into how to use a dichotomous key. Scientists use dichotomous keys all the time to help them identify plants

and animals.

For more information on tree identification, check out a field guide from the library or see this website:

https://treespnw.forestry.oregonstate.edu/

(Optional) Hands-on Science Activities

At another point in the week, the students can do one or more of the following optional activities.

- **Tree Classification Book** – Choose more than one leaf to identify and use those leaves to make a book about the trees in our area. You can download free templates for a book like this from here:

  https://elementalscience.com/products/fall-leaf-book-free-printable
- **Dichotomous Key** – Create your own dichotomous key for a group of items in your room. Have the students follow the key to identify what the items are called.
- **New Species** – Draw your own species of animal, then have the students use the classification chart to assign it a phylum, class, order, family, and genus.

Write: Keeping a Journal

Journaling Sheets

The students should have already recorded what they learned about classification on their notes page (LJ p. 27) and completed the experiment sheet (LJ pp. 24-25). On the notes page, the students can also answer the bonus question. Here is an answer for your reference as you discuss this with your students.

- **Bonus Question** – How has a deeper understanding of DNA changed Linnaeus's classification ideas? (As we learn more about DNA, the science of classifying organisms is changing. This article from Science News for Students is a good one to get the students thinking about how DNA affects classification: https://student.societyforscience.org/article/bird-dna-leads-strange-family-tree.)

Vocabulary

At some point in the week, have the students look up the following words and add the definitions to the unit vocabulary sheet on LJ pp. 14-16:

- **Classification** – A way of identifying or grouping living things.
- **Kingdom** – The highest category into which living things are classified.
- **Species** – A group of living things than can breed together in the wild.

Sketch Assignment

See the next page for the correctly labeled sketch. (LJ p. 26)

Divisions of Life

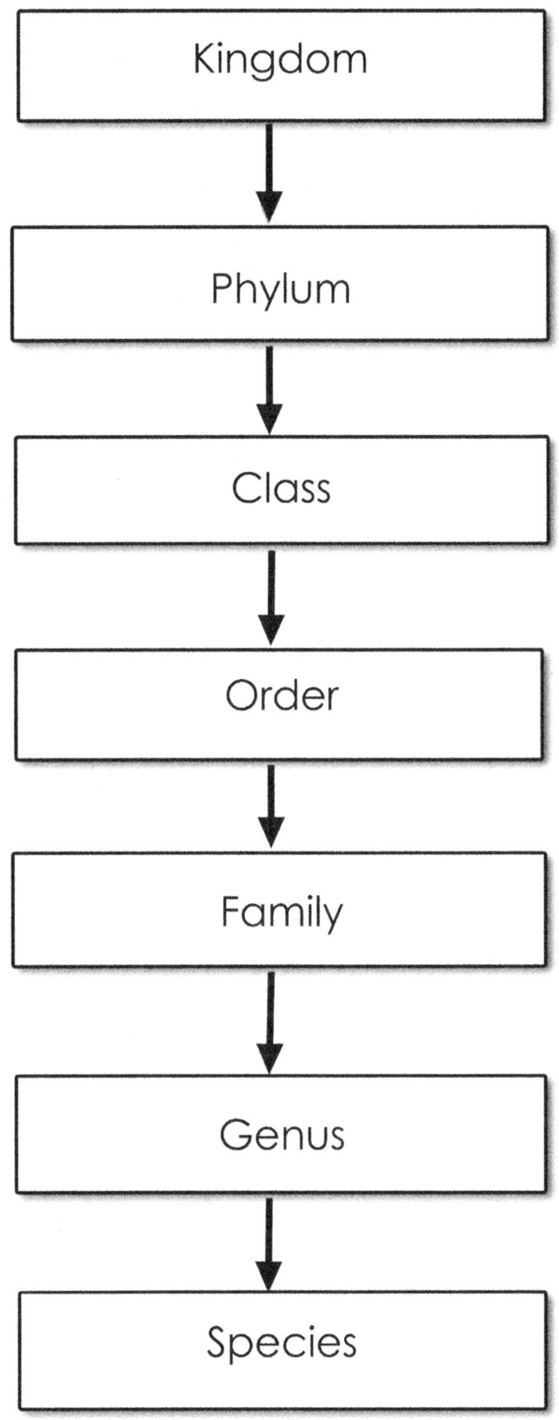

(Optional) Memory Work

At another point in the week, the students can work on the following memory work - the Divisions of Life.

Lesson 3 Grid Schedule

Supplies Needed	
Experiments	• A place to observe, Access to the Internet or a local field guide
Optional Activities	• BBC Planet Earth DVDs • Paper, Colored pencils

Weekly Schedule

	Day 1	Day 2	Day 3	Day 4
Read	❑ Read Chapter 3: "Milo Chuckled in Class" in *The PB Nanner Fluffer Test*	❑ Choose one of the encyclopedia readings to assign and discuss what was read using the questions provided	❑ *(Optional)* Choose another one of the encyclopedia readings to assign	❑ *(Optional)* Choose another one of the encyclopedia readings to assign
Do	❑ Do the "Backyard Animal Observation" experiment		❑ *(Optional)* Watch the BBC Planet Earth DVD Series	❑ *(Optional)* Make a poster of the migration routes discussed in the BBC Planet Earth DVD Series
Write	❑ Fill out the experiment sheet	❑ Write about migration	❑ Color and label the migration sketch ❑ Define hemisphere and migration	❑ *(Optional)* Work on memorizing the Divisions of Life

Lesson 3 List Schedule

Essentials

Read

- O Read Chapter 3: "Milo Chuckled in Class" in *The PB Nanner Fluffer Test.*
- O Choose one of the encyclopedia readings to assign and discuss what was read using the questions provided.

Do

- O Do the "Backyard Animal Observation" experiment. Fill out the lab report as you do the experiment.

Write

- O Write about migration.
- O Look up hemisphere and migration. Add the definitions to the unit vocabulary sheet.
- O Label the migration sketch.

(Optional) Extras

Read

- O Choose another one (or two) of the encyclopedia readings to assign.

Do

- O Watch the BBC Planet Earth DVD Series.
- O Make a poster of the migration routes discussed in the BBC Planet Earth DVD Series.

Write

- O Work on memorizing the Divisions of Life.

Supplies Needed	
Experiments	• A place to observe, Access to the Internet or a local field guide
Optional Activities	• BBC Planet Earth DVDs • Paper, Colored pencils

Lesson 3 Science Focus: Migration

Read: Gathering Information

Key Idea

Animals migrate to breed and to find food.

Week's Readings

First, spark an interest by having the students read this week's story opener:

- Read Chapter 3 in *The PB Nanner Fluffer Test*: "Milo Chuckled in Class." (Your student will hang out with Milo as he goes to co-op class and has lunch with Addison and her friends.)

Then follow the story up with some facts. You can choose which resource works best for your students:

- *Usborne Science Encyclopedia*: "Migration (section)" p. 329
- *Kingfisher Science Encyclopedia*: "Migration" p. 93
- *Usborne Illustrated Dictionary of Science:* "Section on Migration" p. 237
- *Everything You Need to Ace Science in One Big Fat Notebook:* "Section on Migration" p. 370

Discussion Questions

After all that reading, use the following questions to guide your discussion time with the students.

1. Why do animals migrate? (Animals migrate to breed and to find food.)
2. How do animals find their way during migration? (Animals find their way during migration by instinct. They can also navigate using the stars to find their way during migration.)
3. Which animals are the most common migrants? (Birds are the most common migrants.)
4. Name two ways that birds use to navigate. (Birds navigate by following landmarks or geographic features. They also use the Sun, the stars, and the Earth's magnetic field.)

After the discussion time, have the students write a list of facts or a paragraph about what they read. Their writings should include at least a few facts from above.

(Optional) Reading Selections

At another point in the week, the students can check out these optional books:

- *National Geographic Readers: Great Migrations Amazing Animal Journeys* by Laura Marsh
- *When Penguins Cross the Ice (Extraordinary Migrations)* by Sharon Katz Cooper and Tom Leonard
- *Migration: Incredible Animal Journeys* by Mike Unwin

Do: Experimenting with Science

Experiment: Backyard Animal Observations

> Note -
> In an ideal world, you would do this observation with your students so that you can help them identify the animals in the field. But since we all live in reality, we have written the directions as if you are not out with the students. If you can spare the 30 minutes, adjust the procedure as necessary.

Materials

You will need the following items:

- ✓ A place to observe animals
- ✓ Access to the Internet or a local field guide

Introduction

Read the following with your students, while they follow along in their lab journal.

> Milo learned about migration in his co-op class. Often when we think of animals migrating, we think about those creatures that are far off in some distant space that we do not see. But there are animals the migrate through your area! In this experiment, you are going to observe your backyard for animals, then identify those animals, and finally do a bit of research to determine if they migrate.

Directions

Have the students do the following experiment to observe and learn about local fauna (animals) under your supervision.

1. Go outside.
2. Find a quiet place to sit and observe for about 30 minutes.
3. Record your observations on the following chart. (This chart is included on LJ p. 30.)
4. After the time is up, head inside to research the animals you observed using either the Internet or a local field guide. Begin by identifying the animals if you have not done so already. Then, mark down if these animals are seasonal to your area or if they live there year-round.

Results and Explanation

When they are done, read the results and explanation to the students.

> You should be able to observe several animals, identify what they are, and determine if they migrate into the area or not.

Discuss how the results from their experiment looked like or didn't look like the expected results. Then, read the following explanation to the students.

> Again, this experiment was designed to help you practice a key skill you will need as a scientist - your observations skills. It was also meant to give you an opportunity to learn about their local habitat.

(Optional) Hands-on Science Activities

At another point in the week, the students can do one or more of the following optional activities.

- **Video** – Watch the BBC Planet Earth DVD Series.
- **Poster** – Make a poster of the migration routes discussed in the BBC Planet Earth DVD Series.

Write: Keeping a Journal

Journaling Sheets

The students should have already recorded what they learned about migration on their notes page (LJ p. 33) and completed the experiment sheet (LJ pp. 30-31). On the notes page, the students can also answer the bonus question. Here is an answer for your reference as you discuss this with your students.

- **Bonus Question** – If you could migrate, where would you go? Why? (Answers will vary.)

Vocabulary

At some point in the week, have the students look up the following words and add the definitions to the unit vocabulary sheet on LJ pp. 14-16:

- **Hemisphere** – One of the halves of the Earth created by an imaginary division along the equator.
- **Migration** – A journey by an animal to a new habitat.

Sketch

Here is the correctly labeled sketch: (LJ p. 32)

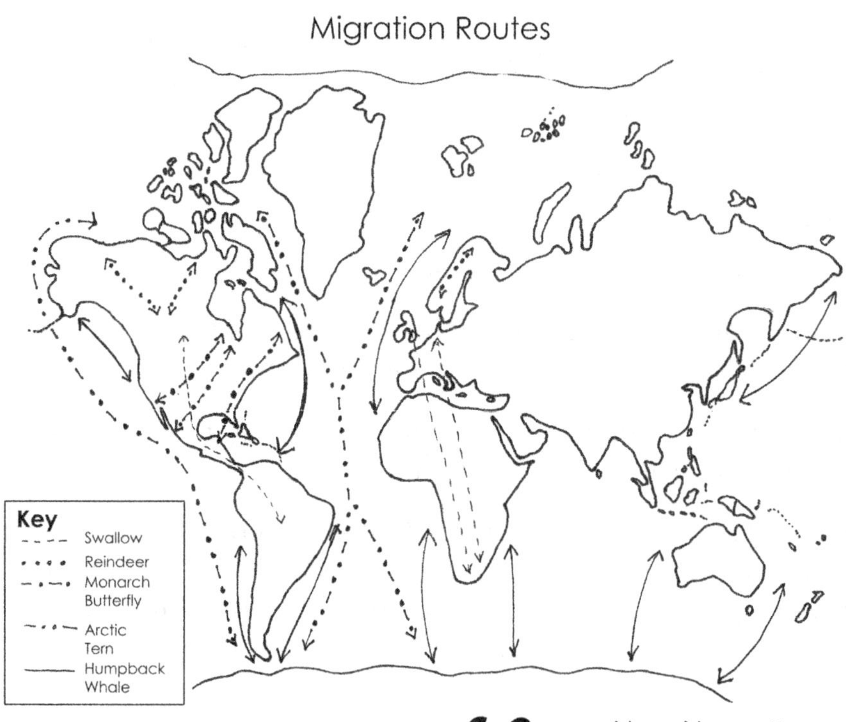

(Optional) Memory Work

At another point in the week, the students can work on the following memory work - the Divisions of Life.

Lesson 4 Grid Schedule

Supplies Needed	
Experiments	• Gloves, Tweezers, Wooden stick (coffee stirrer or popsicle stick will work), Newspaper or paper towel, Owl pellet
Optional Activities	• 20 Food items from pantry • Posterboard, Coloring Pencils

Weekly Schedule

	Day 1	Day 2	Day 3	Day 4
Read	❑ Read Chapter 4: "Milo Gave a Test" in *The PB Nanner Fluffer Test*	❑ Choose one of the encyclopedia readings to assign and discuss what was read using the questions provided	❑ *(Optional)* Choose another one of the encyclopedia readings to assign	❑ *(Optional)* Choose another one of the encyclopedia readings to assign
Do	❑ Do the "Owl Pellet" experiment		❑ *(Optional)* Do the Dietary Selection activity	❑ *(Optional)* Make an Autotroph vs. Heterotroph poster
Write	❑ Fill out the experiment sheet	❑ Write about animal diet ❑ Define autotroph and heterotroph	❑ Color and label the simple food chain sketch ❑ Define carnivore, herbivore, and omnivore	❑ *(Optional)* Work on memorizing the Divisions of Life

Lesson 4 List Schedule

Essentials

Read

- ○ Read Chapter 4: "Milo Gave a Test" in *The PB Nanner Fluffer Test.*
- ○ Choose one of the encyclopedia readings to assign and discuss what was read using the questions provided.

Do

- ○ Do the "Owl Pellet" experiment. Fill out the lab report as you do the experiment.

Write

- ○ Write about animal diet.
- ○ Look up autotroph, heterotroph, carnivore, herbivore, and omnivore. Add the definitions to the unit vocabulary sheet.
- ○ Label the simple food chain sketch.

(Optional) Extras

Read

- ○ Choose another one (or two) of the encyclopedia readings to assign.

Do

- ○ Make an Autotroph vs. Heterotroph poster.
- ○ Do the Dietary Selection activity.

Write

- ○ Work on memorizing the Divisions of Life.

Supplies Needed	
Experiments	• Gloves, Tweezers, Wooden stick (coffee stirrer or popsicle stick will work), Newspaper or paper towel, Owl pellet
Optional Activities	• 20 Food items from pantry • Posterboard, Coloring Pencils

Lesson 4 Science Focus: Animal Diet

Read: Gathering Information

> **Key Idea**
>
> **Animal diet varies and each species is adapted for the food they eat.**

Week's Readings

First, spark an interest by having the students read this week's story opener:

- Read Chapter 4 in *The PB Nanner Fluffer Test*: "Milo Gave a Test." (Your student will read about Milo giving his PB Naner Fluffer test and find out if Addison, Eric, and MJ pass it.)

Then follow the story up with some facts. You can choose which resource works best for your students:

- *Usborne Science Encyclopedia*: "Feeding" pp. 310-311
- *Kingfisher Science Encyclopedia*: "Teeth and Digestion" pp. 312-313, "Food Chains" p. 435
- *Usborne Illustrated Dictionary of Science:* "Within An Ecosystem" pp. 234-235
- *Everything You Need to Ace Science in One Big Fat Notebook:* "Herbivore, Omnivore, and Carnivore" p. 362, "Feeding Relationship, Food Chains, and Food Webs" pp. 487-490

Discussion Questions

After all that reading, use the following questions to guide your discussion time with the students.

1. What is phagocytosis? (Phagocytosis is when a single-celled organism feeds by surrounding, engulfing, and breaking down a particle.)
2. What are the different parts of an insect's mouth and what are they used for? (An insect's mouth has a mandible, which is used for holding and biting, a maxillae, which is used to push food into the mouth, a labrum, which is used to cover and protect the mouthparts, and a labium, which is also used to push food into the mouth.)
3. How do carnivore's teeth differ from an herbivore's teeth? (A carnivore's teeth are sharper so that it can grip and slice through meat. An herbivore's teeth are more rigid, so that it can crush and chew its food.)
4. What can omnivores eat? (Omnivores can eat almost anything edible, both plants and meat, from their environment.)
5. What two adaptations do birds have in their digestive system to deal with solid food? (Birds have a crop that stores food after it has been swallowed and a gizzard that grinds up the whole food.)

After the discussion time, have the students write a list of facts or a paragraph about what they read. Their writings should include at least a few facts from above.

(Optional) Reading Selections

At another point in the week, the students can check out these optional books:

- *Animals Eating: How Animals Chomp, Chew, Slurp and Swallow* by Pamela Hickman

Do: Experimenting with Science

Experiment: Owl Pellet

Materials

You will need the following items:

- ✓ Gloves, Tweezers, Wooden stick (coffee stirrer or popsicle stick will work)
- ✓ Newspaper or paper towel
- ✓ Owl pellet

Introduction

Read the following with your students, while they follow along in their lab journal.

Milo and his friends learn about animal diet, but today we're going to see an animal's diet firsthand in an owl pellet. Unlike other birds, the food goes directly from the owl's mouth into its gizzard where digestion begins. The food is separated into digestible and indigestible parts. The indigestible parts, like the bones, fur, and sand, are compacted together into an oval-shaped pellet, which the owl regurgitates several hours after it eats. Since the owl swallows the prey virtually whole, most of the bones are intact and can be viewed quite easily in the pellet.

Directions

Have the students see the remanents of what owls eat under your supervision.

1. Before you begin, lay out the newspaper or paper towels on your work surface and put on the gloves. Most pellets are sterilized, but you should still use these precautions anyways.
2. Open the owl pellet up and examine the outside. Look for fur, feathers, and other identifiable material.
3. Use the wooden stick to gently break the owl pellet up and observe what you see.
4. Use the tweezers to pick up some of the bones you can see.
5. Examine these bones up close to identify whether they came from a rodent, shrew, mole, and small bird. You may also see remnants of feathers, fur, and bits of sand and gravel. Here is a free Bone Identification Chart from Carolina Biological that you can use to help identify the different bones you find.
   💻 https://www.carolina.com/pdf/activities-articles/Owl_Pellet_Bone_Chart_grid.pdf
6. Write down the bones you find under the observations and results.
7. When you are done with your observations, take off the gloves, roll up your foil with the pellet and tools inside, and throw it all away.

(Optional) Hands-on Science Activities

At another point in the week, the students can do one or more of the following optional activities.

- ✐ **Autotrophs vs. Heterotrophs –** Have the students write a paragraph or create a chart explaining the difference between autotrophs and heterotrophs, including examples of each.

- **Dietary Selection –** Have the students go through your pantry, fridge, or local grocery store to choose 20 food items. Then, have them identify which type of diet it would be a part of - herbivore, carnivore, or omnivore.

Write: Keeping a Journal

Journaling Sheets

The students should have already recorded what they learned about animal diet on their notes page (LJ p. 39) and completed the experiment sheet (LJ pp. 36-37). On the notes page, the students can also answer the bonus question. Here is an answer for your reference as you discuss this with your students.

- **Bonus Question** – Do you think it is easier to survive as an herbivore, a carnivore, or an omnivore? (The students' answers will vary, but they should have some kind of reasoning that includes information about the diet of these different types of animals.)

Vocabulary

At some point in the week, have the students look up the following words and add the definitions to the unit vocabulary sheet on LJ pp. 14-16:

- **Autotroph** – A living thing that makes its own food.
- **Heterotroph** – A living thing that eats other living things.
- **Carnivore** – An animal with specially shaped teeth that feeds mainly on meat.
- **Herbivore** – An animal that eats only plant food.
- **Omnivore** – An animal that eats both plant and animal food.

Sketch

Here is the correctly labeled sketch: (LJ p. 38)

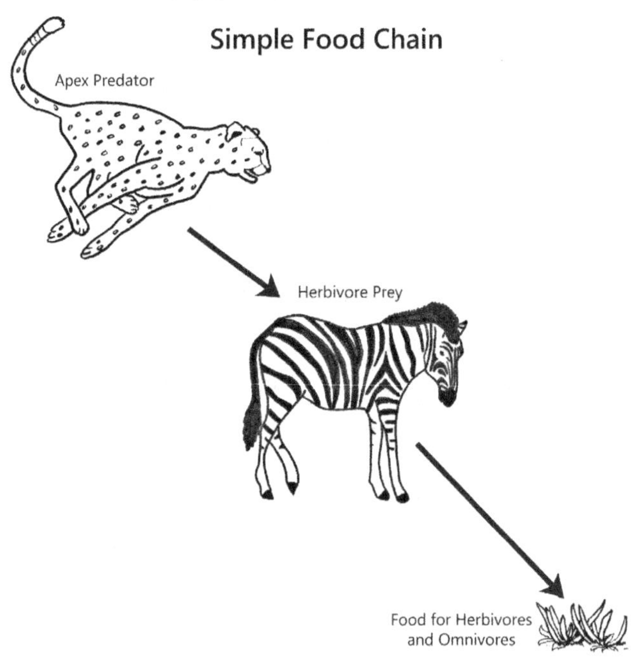

(Optional) Memory Work

At another point in the week, the students can work on the following memory work - the Divisions of Life.

Lesson 5 Grid Schedule

Supplies Needed	
Experiments	• Newspaper, Plain paper, Black and green construction paper
Optional Activities	• Posterboard • Paper, Coloring Pencils

Weekly Schedule

	Day 1	Day 2	Day 3	Day 4
Read	❑ Read Chapter 5: "Milo Accepted His Mission" in *The PB Nanner Fluffer Test*	❑ Choose one of the encyclopedia readings to assign and discuss what was read using the questions provided	❑ *(Optional)* Choose another one of the encyclopedia readings to assign	❑ *(Optional)* Choose another one of the encyclopedia readings to assign
Do	❑ Do the "How does camouflage work?" experiment		❑ *(Optional)* Make a Defense poster	❑ *(Optional)* Do the Create an Animal activity
Write	❑ Fill out the experiment sheet	❑ Write about animal defenses	❑ Color and label the 5 types of animal defenses sketch ❑ Define camouflage, predator, and prey	❑ *(Optional)* Take the Unit 1 Test

Lesson 5 List Schedule

Essentials

Read

- O Read Chapter 5: "Milo Accepted His Mission" in *The PB Nanner Fluffer Test.*
- O Choose one of the encyclopedia readings to assign and discuss what was read using the questions provided.

Do

- O Do the "How does camouflage work?" experiment. Fill out the lab report as you do the experiment.

Write

- O Write about animal defenses and adaptations.
- O Look up camouflage, predator, and prey. Add the definitions to the unit vocabulary sheet.
- O Label the 5 types of animal defenses sketch.

(Optional) Extras

Read

- O Choose another one (or two) of the encyclopedia readings to assign.

Do

- O Make a Defense poster.
- O Create an Animal activity.

Write

- O Take the Unit 1 Test.

Supplies Needed	
Experiments	• Newspaper, Plain paper, Black and green construction paper
Optional Activities	• Posterboard • Paper, Coloring Pencils

Lesson 5 Science Focus: Animal Defense

Read: Gathering Information

Key Idea

Predators hunt for food. Prey (the food) evades by using defenses.

Week's Readings

First, spark an interest by having the students read this week's story opener:

- Read Chapter 5 in *The PB Nanner Fluffer Test*: "Milo Accepted His Mission." (Your student will read about Milo learning about camouflage and learn if he finally accepts his parents' mission.)

Then follow the story up with some facts. You can choose which resource works best for your students:

- *Usborne Science Encyclopedia*: "Sending Messages" pp. 318-319
- *Kingfisher Science Encyclopedia*: "Adaptation and Defense" p. 95
- *Everything You Need to Ace Science in One Big Fat Notebook:* "Relationships Among Populations" pp. 485-487

Discussion Questions

After all that reading, use the following questions to guide your discussion time with the students.

1. What is the difference between adaptation and defense? (An adaptation is a change that an animal or plant makes to better suit the environment it lives in. A defense is the way that an animal or plant protects itself.)
2. What are the five main types of defensive weapons (give examples of each)? (The five main types of defensive weapons are mechanical, chemical, visual, electrical, and behavioral. Horns, spines, thorns, and armor are all examples of mechanical defenses. Substances that sting or poison are examples of chemical defenses. Camouflage and bright colorings are examples of visual defenses. Discharging electricity to stun prey or warn predators is an example of electrical defense. Hiding or running are examples of behavioral defense.)
3. What are the two different types of adaptations? (The two different type of adaptations are genotypic and phenotypic.)
4. What are genotypic adaptations? (Genotypic adaptations are passed down through genes from generation to generation.) Phenotypic adaptations? (Phenotypic adaptations occur within the lifetime of an organism.)

After the discussion time, have the students write a list of facts or a paragraph about what they read. Their writings should include at least a few facts from above.

(Optional) Reading Selections

At another point in the week, the students can check out these optional books:

- *Showdown: Animal Defenses (TIME FOR KIDS®)* by Jennifer Kroll
- *When Lunch Fights Back: Wickedly Clever Animal Defenses* by Rebecca L. Johnson

📖 *How Not to Get Eaten: More than 75 Incredible Animal Defenses (Wonders of Wildlife)* by Josette Reeves and Asia Orlando

Do: Experimenting with Science

Experiment: How does camouflage work?

Note -
This seems like a simplistic experiment and that's because it is. But if you let your students loose and allow them to imagine what would work best, they will discover quite a lot about camouflage all on their own, which is basically the whole point of experiments!

Materials

You will need the following items:

- ✓ Newspaper
- ✓ Plain paper
- ✓ Black and green construction paper

Introduction

Read the following with your students, while they follow along in their lab journal.

> Milo's sketch perfectly depicted the predator/prey relationship. But animals don't like to become a predator's prey, so they have adapted defenses to protect themselves from harm. One of these defenses is to basically hide in plain sight, otherwise known as camouflage. In today's experiment, you are going to use bits of paper to see how this animal defense works.

Directions

Have the students use bits of paper to learn more about how camouflage works under your supervision.

1. Cut out several of the same shapes from the newspaper, plain paper, black construction paper, and green construction paper.
2. Lay out a sheet of newspaper on the table. Then lay the shapes on top of the paper. Observe which shapes are easier to find. (The students can even have a partner come in and try to identify the number and types of shapes that are on the newspaper.)
3. After that, use what you observed and learned in step 2 to cut several different shapes from the paper that was the hardest to see. Again, lay out these shapes to test which is the hardest to find. (Once more, the students can have a partner come in and try to identify the number of shapes that are on the newspaper.)
4. Finally, move the shapes around. Ask yourself, is there a better position or place that will camouflage the shapes more adequately?

Results and Explanation

When they are done, read the results and explanation to the students.

> You should have noticed the green shapes first and the newspaper shapes last. That said, these results with the different shapes will vary. This is because the newspaper is most like the background that it was

on, while the green shapes are the most different.

Discuss how the results from their experiment looked like or didn't look like the expected results. Then, read the following explanation to the students.

> Animals use this same concept to hide themselves in their environment. In biology, we call this being camouflaged. The animal has similar colors to those found in its habitat, which can keep us from noticing it. It also helps the animal hide from predators, which increases its chance of survival.

(Optional) Hands-on Science Activities

At another point in the week, the students can do one or more of the following optional activities.

- **Defense Poster** – Have the students create a poster that depicts each of the different types of defense from the memory work.
- **Create an Animal** – Have the students design their own animal that uses camouflage for defense. Draw a picture of the animal and give a brief description of it.

Write: Keeping a Journal

Journaling Sheets

The students should have already recorded what they learned about adaptation and animal defenses on their notes page (LJ p. 43) and completed the experiment sheet (LJ pp. 40-41).

Vocabulary

At some point in the week, have the students look up the following words and add the definitions to the unit vocabulary sheet on LJ pp. 14-16:

- **Camouflage** – The way animals hide by blending in with their surroundings.
- **Predator** – An animal that kills and eats other animals.
- **Prey** – An animal that is killed and eaten by other animals.

Sketch

See the next page for the correctly labeled sketch. (LJ p. 42)

(Optional) Animal Basics Test

The students can complete the optional unit test found on pp. 145-147 in the Appendix. The answers for this test are found on pp. 46-47. A printable copy of the test can be downloaded for free from:

- https://elementalscience.com/blogs/resources/sq100

5 Types of Animal Defenses

Mechanical defense

Chemical defense

Behavorial defense

Electrical defense

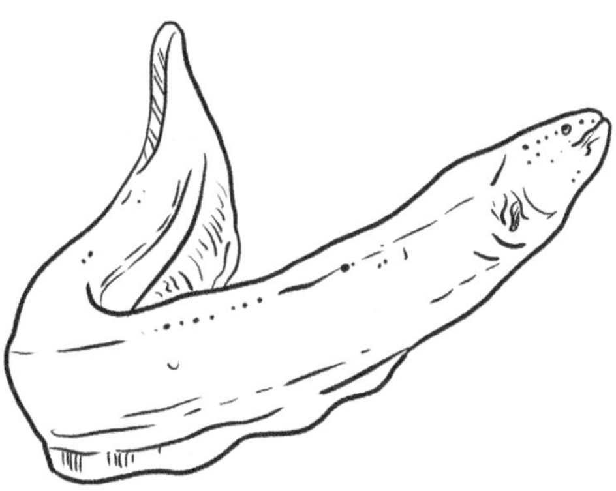

Visual defense

Animal Basics Test Answers

Note -
If you need a digital copy of the unit test, you can download one for free from this program's resource page. Just head to:
https://elementalscience.com/blogs/resources/sq100

Vocabulary Matching (2 points each)

1. H
2. M
3. J
4. A
5. C
6. F
7. K
8. I
9. D
10. G
11. B
12. L
13. N
14. O
15. E

True and False (2 points each)

1. False (*Plant cells have cell walls and chloroplasts, which animal cells do not have.*)
2. True
3. True
4. False (*DNA fingerprinting is one of the tools scientists can use in deciding how to classify a living thing.*)
5. True
6. False (*Birds are the most common migrants.*)
7. True
8. False (*Birds have a crop that stores food after it has been swallowed and a gizzard that grinds up the whole food.*)
9. True
10. True

Short Answer

1. (6 points) The two stages of cell division in a plant are called mitosis and cytokinesis. During mitosis, the nucleus divides, creating two parts; each part becomes a new nucleus. During

cytokinesis, a cell plate forms along the dividing line and a new cell wall builds up along the plate.

2. (7 points) Kingdom, Phylum, Class, Order, Family, Genus, Species
3. (4 points) Birds navigate by following landmarks or geographic features. They also use the Sun, the stars, and the Earth's magnetic field.
4. (4 points) A carnivore's teeth are sharp so that it can grip and slice through meat. An herbivore's teeth are rigid, so that it can crush and chew its food.
5. (4 points) Genotypic adaptations are passed down through genes from generation to generation. Phenotypic adaptations occur within the lifetime of an organism.

Total points: 75

SciQuest

Milo's Mission Teacher Notes

Vertebrates Unit

What you'll study in this unit-

Memory Work-

Basic Classes of Vertebrates (LJ p. 49)

Lesson 6 Grid Schedule

Supplies Needed	
Experiments	• 1 Large clear glass jar or bowl, 3 Small balloons, 3 Small marbles, Ruler, Water
Optional Activities	• Salt, Supplies from the experiment • Goldfish, Fish tank and supplies

Weekly Schedule

	Day 1	Day 2	Day 3	Day 4
Read	❑ Read Chapter 6: "Milo Went on a Field Trip" in *The PB Nanner Fluffer Test*	❑ Choose one of the encyclopedia readings to assign and discuss what was read using the questions provided	❑ *(Optional)* Choose another one of the encyclopedia readings to assign	❑ *(Optional)* Read one or both of the optional reading selections
Do	❑ Do the "Which fish will float higher?" experiment	❑ *(Optional)* Do the Saltwater vs. Freshwater activity	❑ *(Optional)* Set up a goldfish tank	
Write	❑ Fill out the experiment sheet	❑ Write about fish	❑ Color and label the anatomy of a fish sketch ❑ Define gills, swim bladder, and vertebrate	❑ *(Optional)* Work on memorizing the Basic Classes of Vertebrates

Lesson 6 List Schedule

Essentials

Read

- O Read Chapter 6: "Milo Went on a Field Trip" in *The PB Nanner Fluffer Test*.
- O Choose one (or more) of the encyclopedia readings to assign and discuss what was read using the questions provided.

Do

- O Do the "Which fish will float higher?" experiment. Fill out the lab report as you do the experiment.

Write

- O Write about fish.
- O Look up gills, swim bladder, and vertebrate. Add the definitions to the unit vocabulary sheet.
- O Label the anatomy of a fish sketch.

(Optional) Extras

Read

- O Read one or both of the optional reading selections.

Do

- O Do the Saltwater vs. Freshwater activity.
- O Set up a goldfish tank.

Write

- O Work on memorizing the Basic Classes of Vertebrates.

Supplies Needed	
Experiments	• 1 Large clear glass jar or bowl, 3 Small balloons, 3 Small marbles, Ruler, Water
Optional Activities	• Salt, Supplies from the experiment • Goldfish, Fish tank and supplies

Lesson 6 Science Focus: Fish

Read: Gathering Information

> **Key Idea**
> **Fish are cold-blooded animals that live in the water.**

Week's Readings

First, spark an interest by having the students read this week's story opener:

- Read Chapter 6 in *The PB Nanner Fluffer Test*: "Milo Went on a Field Trip." (Your student will join Milo and his co-op class on a field trip to the local aquarium.)

Then follow the story up with some facts. You can choose which resource works best for your students:

- *Usborne Science Encyclopedia*: "Movement in Water" pp. 304-305
- *Kingfisher Science Encyclopedia*: "Fish" pp. 78-79
- *Usborne Illustrated Dictionary of Science:* "Swimmers" p. 269, "Gills" p. 272
- *Everything You Need to Ace Science in One Big Fat Notebook:* "Fish" pp. 356-357

Discussion Questions

After the reading, use the following questions to guide your discussion time with the students.

1. What are the three types of fish and their main characteristics? (The three main types of fish are bony fish, cartilaginous fish, and jawless fish. Cartilaginous fish have skeletons and leathery skin, not scales. Jawless fish have a sucking mouth and are a primitive group of fish. Bony fish have a skeleton that is made of bones and scales covering their bodies.)
2. How do saltwater and freshwater fish get rid of excess water from their bodies? (Saltwater fish lose water through osmosis because the seawater is saltier than their own body fluids. Freshwater get rid of excess water by releasing urine because their body fluids are saltier than the water around them.)
3. How do fish usually reproduce? (Female fish lay the eggs, and then the male fish fertilizes them. The eggs hatch and the babies are generally left to fend for themselves.)
4. What is the lateral line and how does a fish use it? (The lateral line is a fluid-filled tube that runs along the fish's body. A fish uses it to sense changes in water pressure.)
5. Explain how a fish breathes in the water. (A fish breathes by drawing water into its mouth and passing that water through the gills. In the gills, dissolved oxygen from the water is absorbed. Then, water exits the fish's body through slits in the gills.)

After the discussion time, have the students write a list of facts or a paragraph about what they read. Their writings should include at least a few facts from above.

(Optional) Reading Selections

At another point in the week, the students can check out these optional books:

- *Around the Ocean in 80 Fish and other Sea Life* by Helen Scales

Do: Experimenting with Science

Experiment: Which fish will float higher?

Materials

You will need the following items:

- ✓ 1 Large clear glass jar or bowl
- ✓ 3 Small balloons
- ✓ 3 Small marbles
- ✓ Ruler
- ✓ Water

Introduction

Read the following with your students, while they follow along in their lab journal.

> Milo and his friends played a game of "Which Fish is it?" at the aquarium. Fish are a part of the phyla chordata, which means that they have a backbone. They also have a unique organ called a swim bladder. The swim bladder is an organ that helps to control the fish's buoyancy. Buoyancy is the ability to float in a liquid. In this experiment, we are going to look at how the amount of air in the swim bladder (or balloon in our case) affects the buoyancy.

Directions

Have the students do the following experiment to learn about swim bladders and how they help fish under your supervision.

1. Fill the glass cup or bowl with cool water.
2. Label each balloon with #1, #2, or #3 and stretch each balloon so that you can place a marble in it.
3. Tie balloon #1 off with just the marble in it and no air. Blow up balloon #2 a little bit and tie it off. Blow up balloon #3 about halfway and tie it off.
4. Place the balloons gently in the water and let them come to rest (about one minute).
5. Then measure where each one stays and record the measurements. (*Be sure to include the units.*)

Results and Explanation

When they are done, read the results and explanation to the students.

> You should have seen that balloon #3 floats the highest, balloon #2 should be in the middle, and balloon #1 should be at the bottom of the bowl. Be sure that the students recorded the units of measurement they used (i.e., inches (in) or centimeters (cm)).

Discuss how the results from their experiment looked like or didn't look like the expected results. Then, read the following explanation to the students.

The balloon in this experiment represents the swim bladder in a fish. The more air that is contained in it, the higher the fish will be in the water. Most bony fish use their swim bladder to control their buoyancy by inflating (filling it with air) or deflating it (letting air out). Cartilaginous fish, such as sharks, do not contain a swim bladder. Instead, they rely on their wing-like pectoral fins to give them the lift they need to swim in the water.

(Optional) Hands-on Science Activities

At another point in the week, the students can do one or more of the following optional activities.

- **Saltwater vs. Freshwater** – Have the students redo the experiment, only use two glass jars this time. Have them fill one with tap water and one with salt water to see if the results change. The students could also try the experiment again with more than one marble.
- **Goldfish Observation** – Set up a fish tank and purchase a fish so that the students can observe fish behavior daily.

Write: Keeping a Journal

Journaling Sheets

The students should have already recorded what they learned about fish on their notes page (LJ p. 57) and completed the experiment sheet (LJ pp. 54-55). On the notes page, the students can also answer the bonus question. Here is an answer for your reference as you discuss this with your students.

- **Bonus Question** – Why do you think fish lay so many eggs? (Fish lay a lot of eggs because they do not watch over them until they hatch. The quantity of eggs ensures that some will survive until they hatch.)

Vocabulary

At some point in the week, have the students look up the following words and add the definitions to the unit vocabulary sheet on LJ pp. 50-52:

- **Gills** – An organ used to breathe underwater.
- **Swim bladder** – A gas-filled bag that helps a fish to float in water.
- **Vertebrate** – An animal with a backbone.

Sketch Assignment

See the next page for the correctly labeled sketch. (LJ p. 56)

(Optional) Memory Work

At another point in the week, the students can work on the following memory work - the Basic Classes of Vertebrates. For this week, we suggest that you focus on the following:

- **Fish** – Cold-blooded animals that live in water; they are covered with scales and breathe through gills.

Anatomy of a Fish

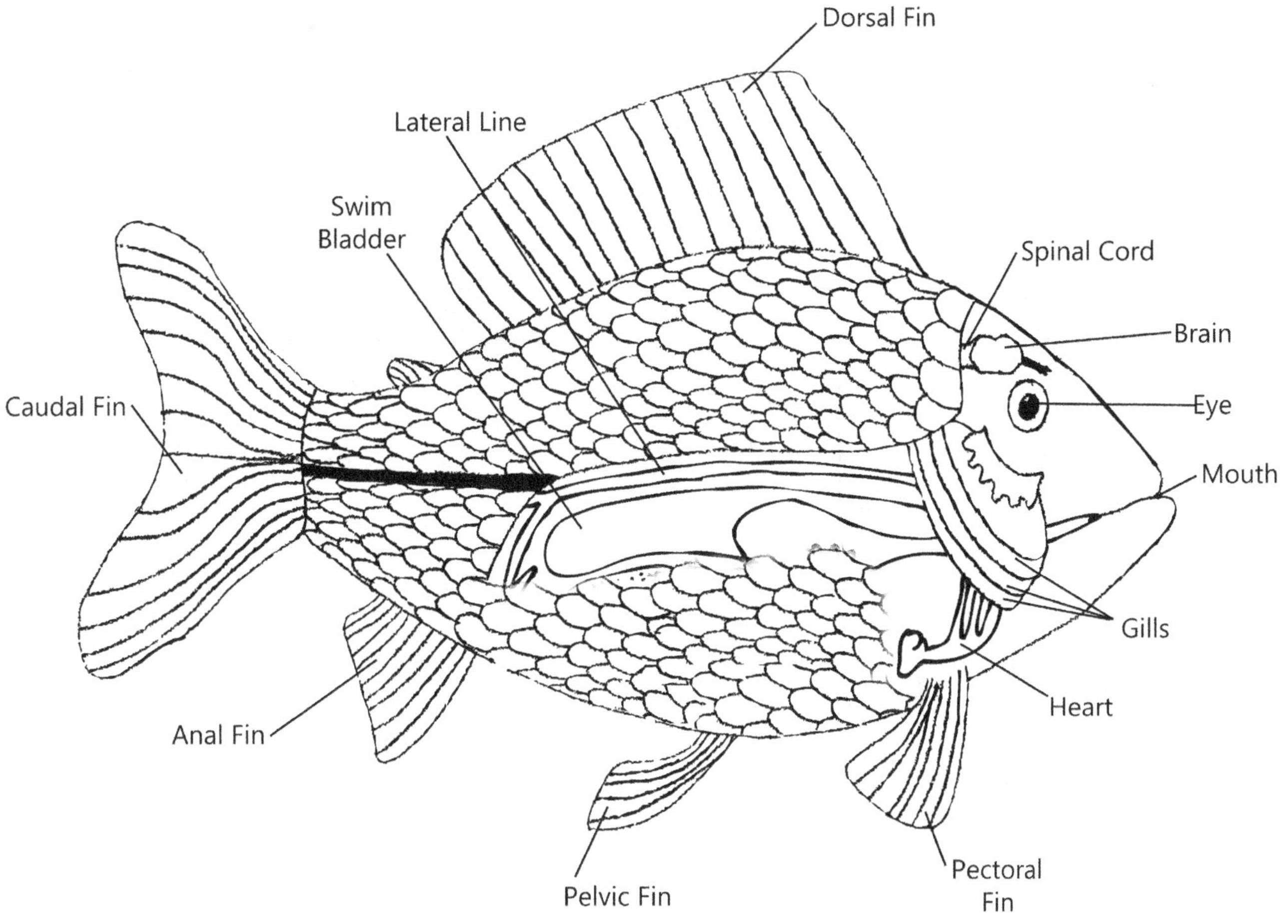

Lesson 7 Grid Schedule

Supplies Needed	
Experiments	• Chopsticks, Tweezers, Pliers, Eye dropper, Sugar, Water, Gummy worms, Peanuts, Seeds, Raisins, Plates
Optional Activities	• Egg • Birdseed, Plate, Bird Identification Chart or Book

Weekly Schedule				
	Day 1	**Day 2**	**Day 3**	**Day 4**
Read	❑ Read Chapter 7: "Milo Ate Watermelon" in *The PB Nanner Fluffer Test*	❑ Choose one of the encyclopedia readings to assign and discuss what was read using the questions provided	❑ *(Optional)* Choose another one of the encyclopedia readings to assign	❑ *(Optional)* Choose another one of the encyclopedia readings to assign
Do	❑ Do the "Which beak?" experiment	❑ *(Optional)* Dissect a Chicken Egg		❑ *(Optional)* Do a Backyard Bird Count
Write	❑ Fill out the experiment sheet	❑ Write about birds	❑ Color and label the three types of bird feet sketch ❑ Define clutch, crop, gizzard, and fledgling	❑ *(Optional)* Work on memorizing the Basic Classes of Vertebrates

Lesson 7 List Schedule

Essentials

Read

- O Read Chapter 7: "Milo Ate Watermelon" in *The PB Nanner Fluffer Test.*
- O Choose one of the encyclopedia readings to assign and discuss what was read using the questions provided.

Do

- O Do the "Which beak?" experiment. Fill out the lab report as you do the experiment.

Write

- O Write about birds.
- O Look up clutch, crop, gizzard, and fledgling. Add the definitions to the unit vocabulary sheet.
- O Label the three types of bird feet sketch.

(Optional) Extras

Read

- O Read one or both of the optional reading selections.

Do

- O Dissect a Chicken Egg.
- O Do a Backyard Bird Count.

Write

- O Work on memorizing the Basic Classes of Vertebrates.

Supplies Needed	
Experiments	• Chopsticks, Tweezers, Pliers, Eye dropper, Sugar, Water, Gummy worms, Peanuts, Seeds, Raisins, Plates
Optional Activities	• Egg • Birdseed, Plate, Bird Identification Chart or Book

Lesson 7 Science Focus: Birds

Read: Gathering Information

> **Key Idea**
>
> **Birds are warm-blooded animals with feathers and wings.**

Week's Readings

First, spark an interest by having the students read this week's story opener:

- Read Chapter 7 in *The PB Nanner Fluffer Test*: "Milo Ate Watermelon." (Your student will attend co-op class with Milo for their experiment day and see what happens when Addison brings Milo watermelon for lunch.)

Then follow the story up with some facts. You can choose which resource works best for your students:

- *Usborne Science Encyclopedia*: "Movement in Air" pp. 306-307
- *Kingfisher Science Encyclopedia*: "Birds" pp. 84-85
- *Usborne Illustrated Dictionary of Science:* "Feathers" p. 267, "Flyers" p. 269
- *Everything You Need to Ace Science in One Big Fat Notebook: :* "Birds" p. 360

Discussion Questions

After all that reading, use the following questions to guide your discussion time with the students.

1. What are three characteristics that enable birds to be so efficient at flight? (Birds are efficient at flight due to their feathers, hollow bones, and strong breast muscles.)
2. What do birds use their feathers for? (Birds use their feathers to fly, to keep warm and dry, and for display.)
3. Why do birds regularly preen? (Birds preen to condition their feathers, so they are ready to fly.)
4. What is the syrinx and what is its purpose? (The syrinx is a structure unique to birds; it is used to make their calls.)
5. How do all birds reproduce? (All birds reproduce by laying eggs.)
6. Name two adaptations that birds have for flight. (Birds have strong breast muscles, well-developed lungs, and a four-chambered heart that work together to help them fly.)

After the discussion time, have the students write a list of facts or a paragraph about what they read. Their writings should include at least a few facts from above.

(Optional) Reading Selections

At another point in the week, the students can check out these optional books:

- *All About Birds: Jaw-dropping Facts About Birds*
- *Bird Trivia: Funny, Strange and Incredible Facts about North American Birds* by Stan Tekiela

Do: Experimenting with Science

Experiment: Which Beak?

Materials

You will need the following items:

- ✓ Chopsticks, Tweezers, Pliers, Eye dropper
- ✓ Sugar, Water, Gummy worms, Peanuts, Seeds, Raisins
- ✓ Plates

Introduction

Read the following with your students, while they follow along in their lab journal.

Milo and his friends did an experiment about bird beaks in their co-op class. And now it's your turn. Remember that all birds have bills or beaks that vary in shape and size. Birds use their beaks to obtain and manipulate their food. They also use their beaks for preening their feathers, for building their nests, and even for climbing! In this experiment, you are going to test to see how the size and shape of a beak relates to the bird's diet.

Directions

Have the students use bits of paper to learn more about how the different types of bird beaks work under your supervision.

1. Match the type of beak to the type of food you think its best suited for picking up.
2. Prepare five plates: one for each of the different types of food.
3. Determine the scale you would like to use when rating ease of pickup. For example, the scale range could be from 1 to 10, with "1" standing for the easiest and "10" standing for the hardest. Fill in the scale range on the experiment sheet.
4. Use the chopsticks to try to pick up each of the different types of food.
5. Rate the ease of pickup on the chart on the experiment sheet. Repeat using the tweezers, pliers, and eye dropper.

Results and Explanation

When they are done, read the results and explanation to the students.

You should see that it was easiest to use:

- ✓ The chopsticks to eat the bread (possibly the gummy worms).
- ✓ The tweezers to eat the seeds and raisins (possibly the gummy worms).
- ✓ The pliers to eat the peanuts.
- ✓ The eye dropper to eat the honey.

Discuss how the results from their experiment looked like or didn't look like the expected results. Then, read the following explanation to the students.

The size and shape of a bird's beak relates to what they eat. The chopsticks represent a long, large bill, like that of a stork whose diet is made up of larger pieces of food such as small fish. The tweezers represent a short, light beak, like that of a finch whose diet is made up of small seeds, berries, and the occasional worm. The pliers represent a short, strong beak like that of a parrot who needs a strong beak to crack seeds and nuts. The eye dropper represents a long, slender bill, like that of a hummingbird whose diet is made up of nectar from flowers.

(Optional) Hands-on Science Activities

At another point in the week, the students can do one or more of the following optional activities.

- **Dissect a Chicken's Egg** – Have the students dissect a chicken egg. They should examine the eggshell, the inner lining, an airspace, the albumin (the white of the egg), and the yolk. On the yolk, the students should see two white cords, which attach the yolk to the inner lining. They may also see a small white dot on the outside of the yolk, which is the beginning of the chicken embryo.
- **Backyard Bird Count** – Have the students observe and record the birds that visit your backyard. You can set out a plate of seeds and keep a bird identification chart or book close by to identify the visitors.

Write: Keeping a Journal

Journaling Sheets

The students should have already recorded what they learned about migration on their notes page (LJ p. 63) and completed the experiment sheet (LJ pp. 60-61). On the notes page, the students can also answer the bonus question. Here is an answer for your reference as you discuss this with your students.

- **Bonus Question** – If you had to make a bird's nest to protect your eggs, what materials would you use and where would you put the nest? (Answers will vary.)

Vocabulary

At some point in the week, have the students look up the following words and add the definitions to the unit vocabulary sheet on LJ pp. 50-52:

- **Clutch** – A set of eggs.
- **Crop** – A pouch in a bird's digestive system where swallowed food is stored.
- **Gizzard** – A chamber in an animal's stomach that grinds up food.
- **Fledgling** – A young bird at the time it leaves the nest.

Sketch

See the next page for the correctly labeled sketch. (LJ p. 62)

(Optional) Memory Work

Three Types of Bird Feet

Hooked claws and talons from a bird of prey.

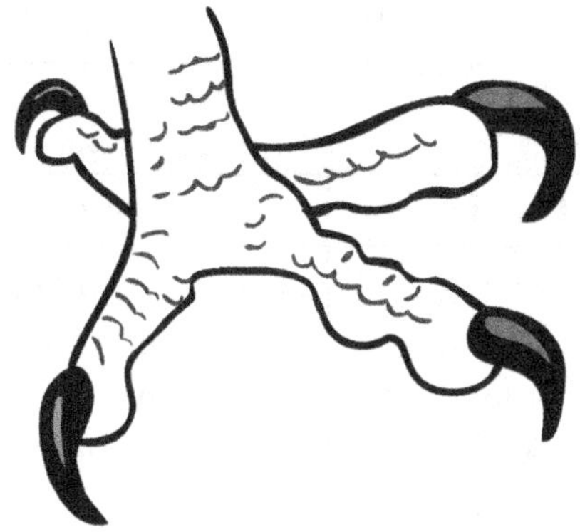

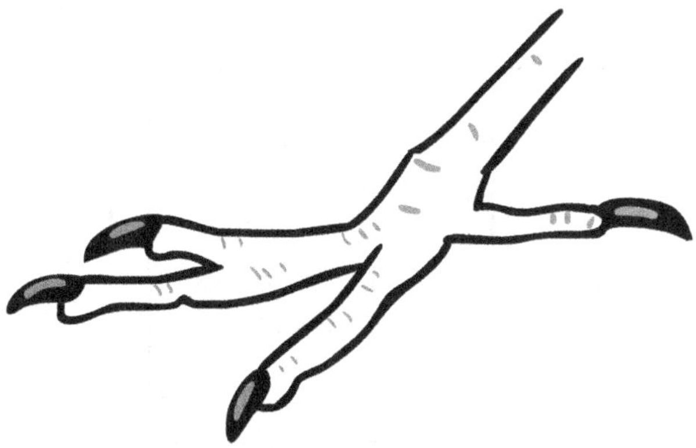

A backward-pointing toe from a perching bird.

Webbed feet from a swimming bird.

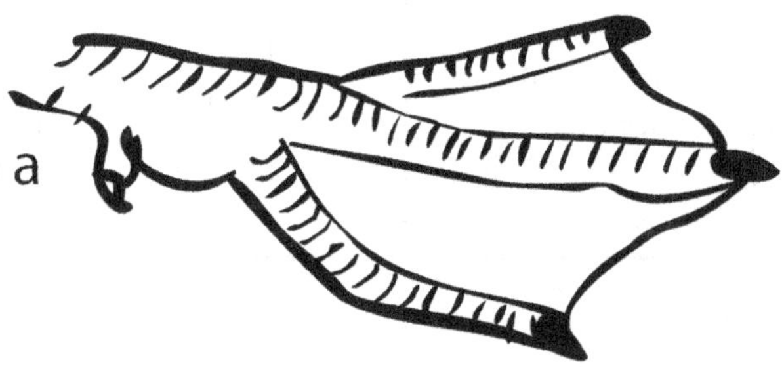

At another point in the week, the students can work on the following memory work - the Basic Classes of Vertebrates. For this week, we suggest that you focus on the following:

- **Birds** – Warm-blooded animals that lay eggs; they have feathers and wings.

Lesson 8 Grid Schedule

Supplies Needed	
Experiments	• 1 Frog dissection kit, 1 Preserved frog
Optional Activities	• Grow-a-frog Kit

Weekly Schedule				
	Day 1	Day 2	Day 3	Day 4
Read	❑ Read Chapter 8: "Milo Walked in the River" in *The PB Nanner Fluffer Test*	❑ Choose one of the encyclopedia readings to assign and discuss what was read using the questions provided	❑ *(Optional)* Choose another one of the encyclopedia readings to assign	❑ *(Optional)* Choose another one of the encyclopedia readings to assign
Do	❑ Do the "Frog Dissection" experiment		❑ *(Optional)* Research and write a report on idicator species	❑ *(Optional)* Set up a Grow A Frog project
Write	❑ Define amphibian and tadpole	❑ Write about amphibians	❑ Color and label the life cycle of a frog	❑ *(Optional)* Work on memorizing the Basic Classes of Vertebrates

Lesson 8 List Schedule

Essentials

Read

- O Read Chapter 8: "Milo Walked in the River" in *The PB Nanner Fluffer Test.*
- O Choose one of the encyclopedia readings to assign and discuss what was read using the questions provided.

Do

- O Do the "Frog Dissection" experiment.

Write

- O Write about amphibians.
- O Look up amphibian and tadpole. Add the definitions to the unit vocabulary sheet.
- O Label the life cycle of a frog sketch.

(Optional) Extras

Read

- O Choose another one (or two) of the encyclopedia readings to assign.

Do

- O Research and write a report on idicator species.
- O Set up a Grow A Frog project.

Write

- O Work on memorizing the Basic Classes of Vertebrates.

Supplies Needed	
Experiments	• 1 Frog dissection kit, 1 Preserved frog
Optional Activities	• Grow-a-frog Kit

Lesson 8 Science Focus: Amphibians

Read: Gathering Information

Key Idea

Amphibians are cold-blooded animals that live on land and in water.

Week's Readings

First, spark an interest by having the students read this week's story opener:

- Read Chapter 8 in *The PB Nanner Fluffer Test*: "Milo Walked in the River." (Your student will go with Milo as he visits the local river with Addison and her family.)

Then follow the story up with some facts. You can choose which resource works best for your students:

- *Usborne Science Encyclopedia*: "Amphibians" pp. 80-81
- *Kingfisher Science Encyclopedia*: "Life Cycles" pp. 328-329
- *Everything You Need to Ace Science in One Big Fat Notebook:* "Amphibians" p. 358

Discussion Questions

After all that reading, use the following questions to guide your discussion time with the students.

1. What are three characteristics of amphibians? (Amphibians are the smallest group of vertebrates. Amphibians are cold-blooded. Amphibians begin life in the water breathing through gills. As they mature, they grow lungs and legs so they can live on land.)
2. What are the three main groups of amphibians? (The three main groups of amphibians are caecilians, frogs/toads, and newts/salamanders.)
3. How can you tell the difference between frogs and toads? (In general, frogs have smoother skin and spend more time in the water than toads do.)
4. Briefly describe the typical amphibian life cycle. (First, embryos develop inside eggs. They hatch as either a tadpole or a larva. The tadpole or larva undergoes metamorphosis to develop into an adult. The adult matures, mates, and lays eggs so that the cycle can begin again.)

After the discussion time, have the students write a list of facts or a paragraph about what they read. Their writings should include at least a few facts from above.

(Optional) Reading Selections

At another point in the week, the students can check out these optional books:

- *Frogs of the World: A Guide to Every Family* by Mark O'Shea
- *Everything You Need to Know About Frogs and Other Slippery Creatures* by DK
- *Reptiles and Amphibians of the Americas: Reptiles and Amphibians of the Americas* by Marty Crump and Andy Charrier

Do: Experimenting with Science

Experiment: Frog Dissection

Materials

You will need the following items:

- ✓ 1 Frog dissection kit
- ✓ 1 Preserved frog

Directions

Have the students do the following experiment to use a dissection to learn more about animal anatomy under your supervision.

1. Please follow all the directions included in the dissection kit. There is no experiment write-up for this week.

> **Alternative Option-**
> If you do not wish to dissect a live frog, you can purchase a good virtual dissection app from Frog Outs:
>
> 💻 https://www.frogvirtualdissection.com/
>
> Or watch a virtual dissction here:
>
> 💻 https://gizmos.explorelearning.com/find-gizmos/lesson-info?resourceId=1097

The purpose of the lab this week is for the students to get some experience with dissection. If you did not purchase the coordinating experiment kit for this program, you can purchase a basic dissection kit from Carolina Biological Supply. There is also a Young Scientist Frog dissection kit that is available from Amazon or Home Science Tools. All these kits should come with instructions on how to dissect the frog.

(Optional) Hands-on Science Activities

At another point in the week, the students can do one or more of the following optional activities.

- **Life Cycle of a Frog** – Raise a tadpole to a frog using a Grow-a-frog kit that you purchase from a science supply store.
- **Indicator Species Report** – Have the students research the potential indicator species for the ecosystem you live in. Have them write a paragraph or create a poster showing what the animals and plants are, and how their presence or absence affects the place they are living in.

Write: Keeping a Journal

Journaling Sheets

The students should have already recorded what they learned about amphibians on their notes page (LJ p. 67). On the notes page, the students can also answer the bonus question. Here is an answer for your reference as you discuss this with your students.

- **Bonus Question** – Why might amphibians be known as "indicator species" for the health of an ecosystem? (Amphibians are considered indicator species for the health of an ecosystem because they live in two different environments over their life cycle. Plus, amphibians also have highly permeable skin.)

Note – An indicated species is a plant or animal whose presence or absence can tell scientists about the health of the environment.

Vocabulary

At some point in the week, have the students look up the following words and add the definitions to the unit vocabulary sheet on LJ pp. 50-52:

- **Amphibian** – A cold-blooded vertebrate that lives partly in water and partly on land.
- **Tadpole** – The immature form of a frog or toad.

Sketch Assignment

Here is the correctly labeled sketch: (LJ p. 66)

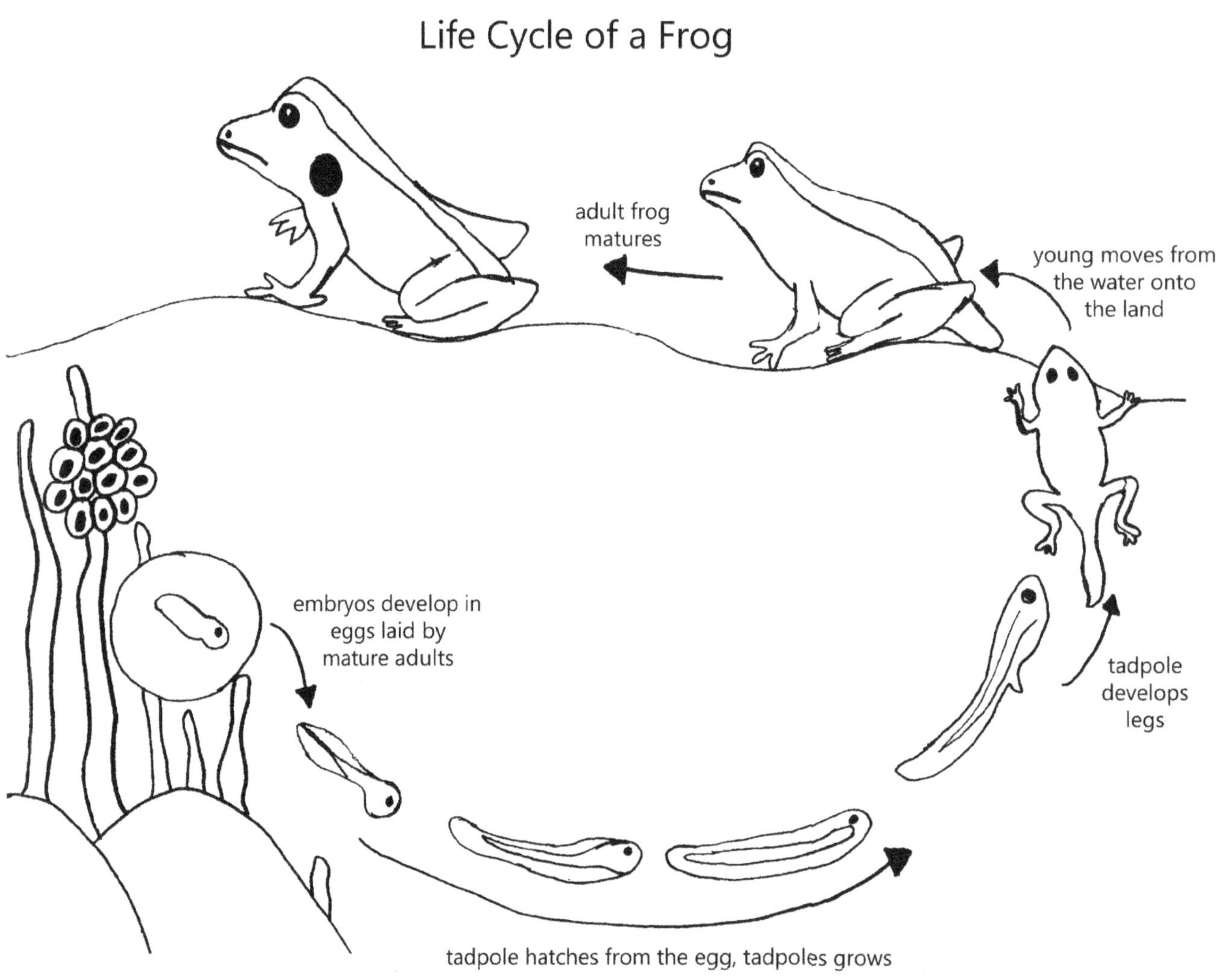

(Optional) Memory Work

At another point in the week, the students can work on the following memory work - the Basic Classes of Vertebrates. For this week, we suggest that you focus on the following:

- **Amphibians** – Cold-blooded animals that live on land and in the water; they have soft skin.

Lesson 9 Grid Schedule

Supplies Needed	
Experiments	• Clothespin, Blindfold, 5 Plates, 5 Pieces of bread with different edible spreads on them (such as garlic, cinnamon sugar, butter, peanut butter, and plain)
Optional Activities	• Glass, Vinegar, Egg, Plastic wrap, Rubber band

Weekly Schedule

	Day 1	**Day 2**	**Day 3**	**Day 4**
Read	❑ Read Chapter 9: "Milo Faced His PE Nemesis" in *The PB Nanner Fluffer Test*	❑ Choose one of the encyclopedia readings to assign and discuss what was read using the questions provided	❑ *(Optional)* Choose another one of the encyclopedia readings to assign	❑ *(Optional)* Choose another one of the encyclopedia readings to assign
Do	❑ Do the "Do I use only my tongue to taste things?" experiment		❑ *(Optional)* Make Reptile Eggs	❑ *(Optional)* Compare Amphibians and Reptiles
Write	❑ Fill out the experiment sheet	❑ Write about reptiles ❑ Define ectotherm, reptile, and scales	❑ Color and label the chameleon defense sketch ❑ Define carnivore, herbivore, and omnivore	❑ *(Optional)* Work on memorizing the Basic Classes of Vertebrates

Lesson 9 List Schedule

Essentials

Read

- O Read Chapter 9: "Milo Faced His PE Nemesis" in *The PB Nanner Fluffer Test.*
- O Choose one of the encyclopedia readings to assign and discuss what was read using the questions provided.

Do

- O Do the "Do I use only my tongue to taste things?" experiment. Fill out the lab report as you do the experiment.

Write

- O Write about reptiles.
- O Look up ectotherm, reptile, and scales. Add the definitions to the unit vocabulary sheet.
- O Label the chameleon defense sketch.

(Optional) Extras

Read

- O Choose another one (or two) of the encyclopedia readings to assign.

Do

- O Make Reptile Eggs.
- O Compare Amphibians and Reptiles.

Write

- O Work on memorizing the Basic Classes of Vertebrates.

Supplies Needed	
Experiments	• Clothespin, Blindfold, 5 Plates, 5 Pieces of bread with different edible spreads on them (such as garlic, cinnamon sugar, butter, peanut butter, and plain)
Optional Activities	• Glass, Vinegar, Egg, Plastic wrap, Rubber band

Lesson 9 Science Focus: Reptiles

Read: Gathering Information

> **Key Idea**
>
> **Reptiles are cold-blooded animals covered with scales.**

Week's Readings

First, spark an interest by having the students read this week's story opener:

- Read Chapter 9 in *The PB Nanner Fluffer Test*: "Milo Faced His PE Nemesis." (Your student will read about Milo and Addison remembering a reptile man visit and overcoming his fear.)

Then follow the story up with some facts. You can choose which resource works best for your students:

- *Usborne Science Encyclopedia*: "Body Covering" pp. 302-303
- *Kingfisher Science Encyclopedia*: "Reptiles" pp. 82-83
- *Everything You Need to Ace Science in One Big Fat Notebook:* "Reptiles" p. 359

Discussion Questions

After all that reading, use the following questions to guide your discussion time with the students.

1. What are some characteristics of a reptile? (Reptiles are cold-blooded animals with dry, scaly skin. Most reptiles lay eggs.)
2. What are the four main groups of reptiles? (The four main groups of reptiles are crocodilians, tortoises and turtles, snakes, and lizards.)
3. How do reptiles control their body temperature? (Reptiles control their body temperature by moving between warm and cool places.)
4. What are some characteristics of crocodilians? (Crocodilians are large carnivores. They have strong jaws and powerful tails. Crocodiles lay eggs on land and are known to fiercely guard their nests.)
5. What characteristic is unique to tortoises and turtles? (Tortoises and turtles have a hard shell that covers most of their bodies. The reptile's head, legs, and tail are exposed, but it can pull these inside the shell if the animal becomes alarmed.)
6. What are some characteristics of snakes? (Snakes are carnivorous reptiles that move by wriggling their bodies.)
7. What do some snakes and lizards do that most reptiles do not? (Some snakes and lizards give birth to live young.)

After the discussion time, have the students write a list of facts or a paragraph about what they read. Their writings should include at least a few facts from above.

(Optional) Reading Selections

At another point in the week, the students can check out these optional books:

- *National Geographic Kids Everything Reptiles* by Blake Hoena

- 🕮 *Reptiles and Amphibians of the Americas: Reptiles and Amphibians of the Americas* by Marty Crump and Andy Charrier
- 🕮 *Discovering Reptiles: The Ultimate Handbook to the Reptiles of the World!* by Thomas Nelson

Do: Experimenting with Science

Experiment: Do I use only my tongue to taste things?

Materials

You will need the following items:

- ✓ Clothespin
- ✓ Blindfold
- ✓ 5 Plates
- ✓ 5 Pieces of bread with different edible spreads on them (such as garlic, cinnamon sugar, butter, peanut butter, and plain)

Introduction

Read the following with your students, while they follow along in their lab journal.

> Milo's friend Addison once touched the scales of a reptile while watching it flick its tongue in and out. Reptiles use their sense of sight, smell, and hearing to find their food. Snakes and some lizards have a special sense organ on the roof of their mouth called the Jacobson's organ. This organ gives them the ability to taste and smell chemical particles in the air with their tongue. In this experiment, you will test to see if we use only our tongues to taste our food.

Directions

Have the students test how our senses help them taste food under your supervision.

1. Prepare the bread with the various edible spreads on five different plates and number the samples from #1-5.
2. Use the clothespin to close your nose. Then have your partner blindfold you and mix up the plates.
3. Have your partner feed each food to you, while you try to guess what each one is.
4. Remove the clothespin. Then taste each of the samples again.
5. Record the results.

Results and Explanation

When they are done, read the results and explanation to the students.

> You should see that you were able to identify food better when the clothespins were off of your nose.

Discuss how the results from their experiment looked like or didn't look like the expected results. Then, read the following explanation to the students.

Our tongues play a major role in how we taste things. Unlike snakes and lizards, we do not have a specialized sense of taste. We need the help of our noses to taste our food because we don't have a Jacobson's organ on the roof of our mouths.

(Optional) Hands-on Science Activities

At another point in the week, the students can do one or more of the following optional activities.

- **Reptile Eggs** – Have the students recreate a reptile egg. You will need a glass, vinegar, egg, plastic wrap, and a rubber band. Have the students place the egg in the glass and slowly pour the vinegar over the egg until the egg is completely covered. Cover the glass with the plastic wrap and secure it with a rubber band. Set the jar aside in a safe place where it can sit undisturbed for 24 hours. After a full day, take off the lid, and use a spoon to fish out the egg. Rinse it gently, and then let the students touch and feel the egg. The calcium in the shell will dissolve and the egg will feel leathery, just like a reptile's egg. Be careful though because it can easily break!
- **Amphibians and Reptiles** – Have the students create a Venn Diagram showing the differences and similarities between amphibians and reptiles.

Write: Keeping a Journal

Journaling Sheets

The students should have already recorded what they learned about reptiles on their notes page (LJ p. 73) and completed the experiment sheet (LJ pp. 70-71). On the notes page, the students can also answer the bonus question. Here is an answer for your reference as you discuss this with your students.

- **Bonus Question** – How do you think being cold-blooded changes what an animal does during the day? (The students' answers will vary, but they should have included something about how reptiles will seek out hotspots and will move slower when cold.)

Vocabulary

At some point in the week, have the students look up the following words and add the definitions to the unit vocabulary sheet on LJ pp. 50-52:

- **Ectotherm** – An animal whose temperature varies with its surroundings; ectothermic animals are also called cold-blooded.
- **Reptile** – A cold-blooded vertebrate with scaly skin.
- **Scales** – Small, overlapping plates that protect the skin.

Sketch

See the next page for the correctly labeled sketch. (LJ p. 72)

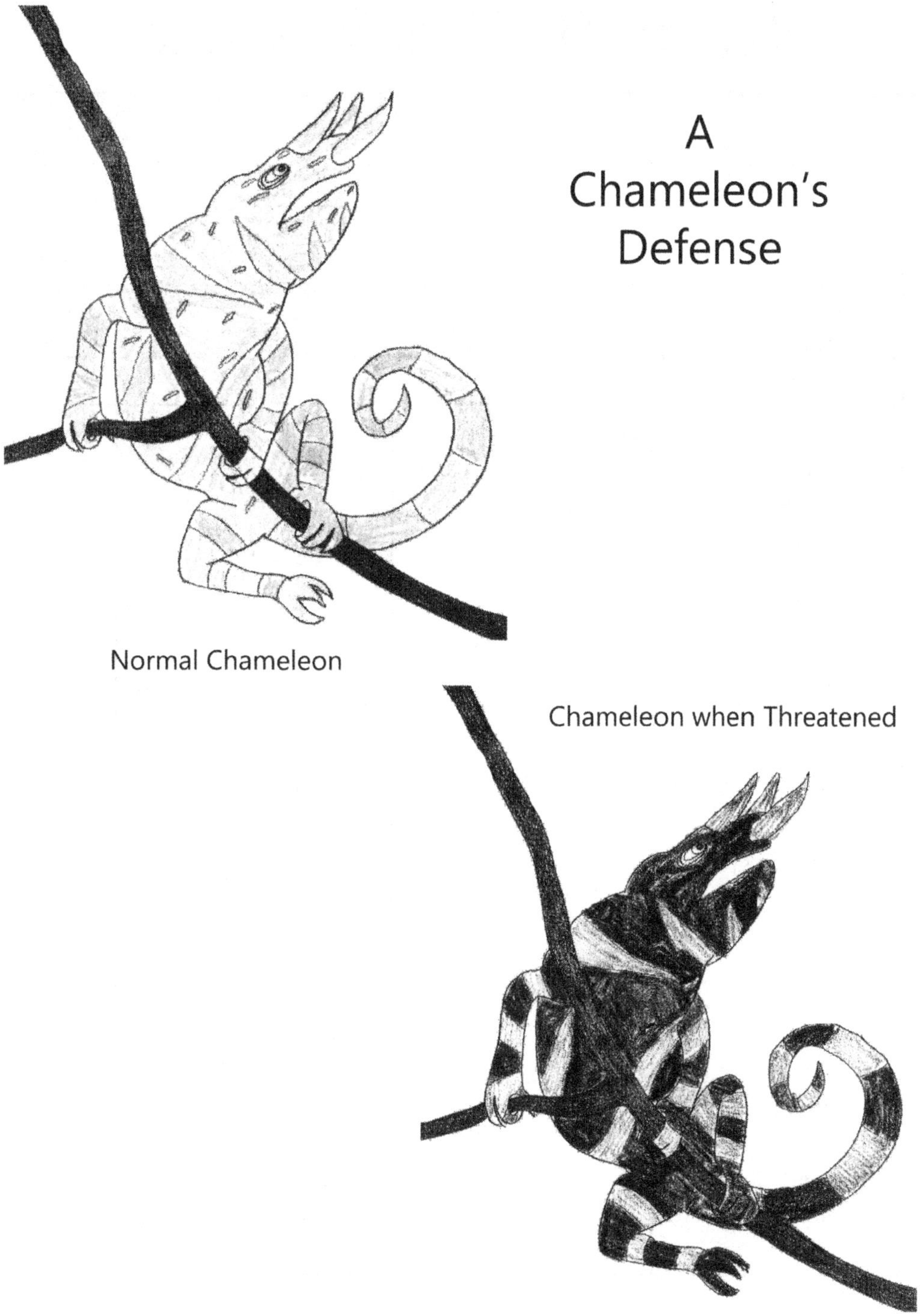

(Optional) Memory Work

At another point in the week, the students can work on the following memory work - the Basic Classes of Vertebrates. For this week, we suggest that you focus on the following:

- **Reptiles** – Cold-blooded animals that lay eggs; they are covered with scales.

Lesson 10 Grid Schedule

Supplies Needed	
Experiments	• Felt, Cotton balls, Water, 1 Large cup, 4 Small cups, Instant read thermometer
Optional Activities	• Paper

Weekly Schedule				
	Day 1	**Day 2**	**Day 3**	**Day 4**
Read	❑ Read Chapter 10: "Milo Navigated a Corn Maze" in *The PB Nanner Fluffer Test*	❑ Choose one of the encyclopedia readings to assign and discuss what was read using the questions provided	❑ *(Optional)* Choose another one of the encyclopedia readings to assign	❑ *(Optional)* Choose another one of the encyclopedia readings to assign
Do	❑ Do the "Which type of fur keeps you warmer?" experiment	❑ *(Optional)* Play with Online Mitosis	❑ *(Optional)* Make a chart showing Mitosis vs. Meiosis	
Write	❑ Fill out the experiment sheet	❑ Write about mammals	❑ Color and label the mitosis sketch ❑ Define endotherm, mammals, mammary gland, and monotreme	❑ *(Optional)* Take the Unit 2 Test

Lesson 10 List Schedule

Essentials

Read

- O Read Chapter 10: "Milo Navigated a Corn Maze" in *The PB Nanner Fluffer Test.*
- O Choose one of the encyclopedia readings to assign and discuss what was read using the questions provided.

Do

- O Do the "Which type of fur keeps you warmer?" experiment. Fill out the lab report as you do the experiment.

Write

- O Write about mammals.
- O Look up endotherm, mammals, mammary gland, and monotreme. Add the definitions to the unit vocabulary sheet.
- O Label the mitosis sketch.

(Optional) Extras

Read

- O Choose another one (or two) of the encyclopedia readings to assign.

Do

- O Play with Online Mitosis.
- O Make a chart showing Mitosis vs. Meiosis.

Write

- O Take the Unit 2 Test.

Supplies Needed	
Experiments	• Felt, Cotton balls, Water, 1 Large cup, 4 Small cups, Instant read thermometer
Optional Activities	• Paper

Lesson 10 Science Focus: Mammals

Read: Gathering Information

> **Key Idea**
>
> **Mammals are warm-blooded animals with fur.**

Week's Readings

First, spark an interest by having the students read this week's story opener:

- Read Chapter 10 in *The PB Nanner Fluffer Test*: "Milo Navigated a Corn Maze." (Your student will read about Milo and his co-ops fall festival, including bunnies!)

Then follow the story up with some facts. You can choose which resource works best for your students:

- *Usborne Science Encyclopedia*: "Movement on Land" pp. 308-309
- *Kingfisher Science Encyclopedia*: "Mammals" pp. 86-87, "Animal Reproduction" pp. 88-89
- *Usborne Illustrated Dictionary of Science:* "Mitosis (cell division)" pp. 240-241, "Animal Reproduction and Development" pp. 276-277
- *Everything You Need to Ace Science in One Big Fat Notebook:* "Cell Division and Mitosis" pp. 321-323, "Mammals" pp. 360-361

Discussion Questions

After all that reading, use the following questions to guide your discussion time with the students.

1. What are two other characteristics of most mammals? (Most mammals are warm-blooded and they feed their young from milk-producing glands.)
2. What are the three groups of mammals? (The three groups of mammals are placental, marsupial, and monotreme.)
3. Explain how the three groups of mammals differ in the way that they reproduce. (In placental mammals, the female's placenta provides all the nutrients for the growing baby. They give birth to fully formed young. In contrast, marsupial mammals give birth to poorly developed young that find their way into a pouch for the remainder of the development. Monotremes lay eggs, unlike any other mammals.)
4. What two factors have contributed to the success of mammals? (Mammals can contribute their survival success to parenting and highly developed senses.)
5. How is animal's lifespan related to growth and reproduction? (The longer an animal takes to mature, mate, and reproduce, the longer its expected lifespan.)
6. Which group of chordates takes the longest to reproduce? (Mammals take the longest to reproduce.) Why? (Mammals take the longest to reproduce because their young generally need months or even years to fully mature.)
7. What are clones? (Clones are genetically identical organisms.)

After the discussion time, have the students write a list of facts or a paragraph about what they read. Their writings should include at least a few facts from above.

(Optional) Reading Selections

At another point in the week, the students can check out these optional books:

- *National Geographic Kids Ultimate Mammalpedia: The Most Complete Mammal Reference Ever* by Stephanie Warren Drimmer

Do: Experimenting with Science

Experiment: Which type of fur keeps you warmer?

Materials

You will need the following items:

- ✓ Felt
- ✓ Cotton balls
- ✓ Water
- ✓ 1 Large cup
- ✓ 4 Small cups
- ✓ Instant read thermometer
- ✓ Foil

Introduction

Read the following with your students, while they follow along in their lab journal.

> Milo and his friends stroked the soft fur of a bunny at the fall festival, but fur is more than something soft to pet. Remember that all mammals are warm-blooded, which means that they are able to maintain a constant internal body temperature no matter what the outside temperature is. Mammals use their food to make energy that helps keep them warm. They also have fur to serve as a barrier to cold air and water. In this experiment, we will look at several types of fur to see which does the best job of keeping the cold out.

Directions

Have the students test to see what materials hold in heat.

1. Line the outside of cup #1 with felt and fill three-quarters of the way with water and label it.
2. Next, take the large cup and place a few cotton balls at the bottom of the cup.
3. Then place cup #2 on top of the cotton balls, and fill in the space with more cotton balls.
4. Then fill cup #2 three-quarters of the way with water and label it.
5. Glue a layer of cotton ball onto a layer of felt and wrap it around cup #3, and then cover it with another layer of felt. Fill cup #3 three-quarters of the way with water and label it.
6. Fill cup #4 three-quarters of the way with water and label it. This is the control.
7. Take the initial temperature of all four cups, record, and cover with foil.
8. Place all four cups into the freezer.
9. Measure and record the temperatures every thirty minutes over a period of three hours.
10. Draw conclusions.

Results and Explanation

When they are done, read the results and explanation to the students.

> You should have seen that cup #3 stays the warmest over the three hours, followed by cup #2, then cup #1, and finally cup #4.

Discuss how the results from their experiment looked like or didn't look like the expected results. Then, read the following explanation to the students.

> The varying thicknesses of fur help to keep different mammals warm. The felt alone (cup #1) simulates a thin layer of fur, like that of a mouse or possum. The cotton balls alone (cup #2) simulate thick fur, like that of a rabbit or fox. The felt and cotton balls (cup #3) simulate a very thick coat of fur, like that of an arctic mammal. Cup #4 represents an animal with no fur and is the control because it will change immediately with the temperature.

(Optional) Hands-on Science Activities

At another point in the week, the students can do one or more of the following optional activities.

- **Online Mitosis –** Watch the following video on mitosis from McGraw Hill:
  https://youtu.be/sU3RFS_QkU8?si=xGZ0ruKVzf5GB2F4
- **Mitosis vs. Meiosis –** Have the students create a chart comparing the differences and similarities between mitosis and meiosis. Use the following article to help:
  http://www.ducksters.com/science/biology/cell_division.php

Write: Keeping a Journal

Journaling Sheets

The students should have already recorded what they learned about mammals on their notes page (LJ p. 79) and completed the experiment sheet (LJ p. 76-77).

Vocabulary

At some point in the week, have the students look up the following words and add the definitions to the unit vocabulary sheet on LJ pp. 50-52:

- **Endotherm –** An animal with a constant temperature; endothermic animals are also called warm-blooded.
- **Mammals –** A warm-blooded animal with hair that feeds its young on milk.
- **Mammary gland –** The milk-producing organ of a female mammal.
- **Monotreme –** A mammal that lays eggs, such as a duck-billed platypus.

Sketch

Here is the correctly labeled sketch: (LJ p. 78)

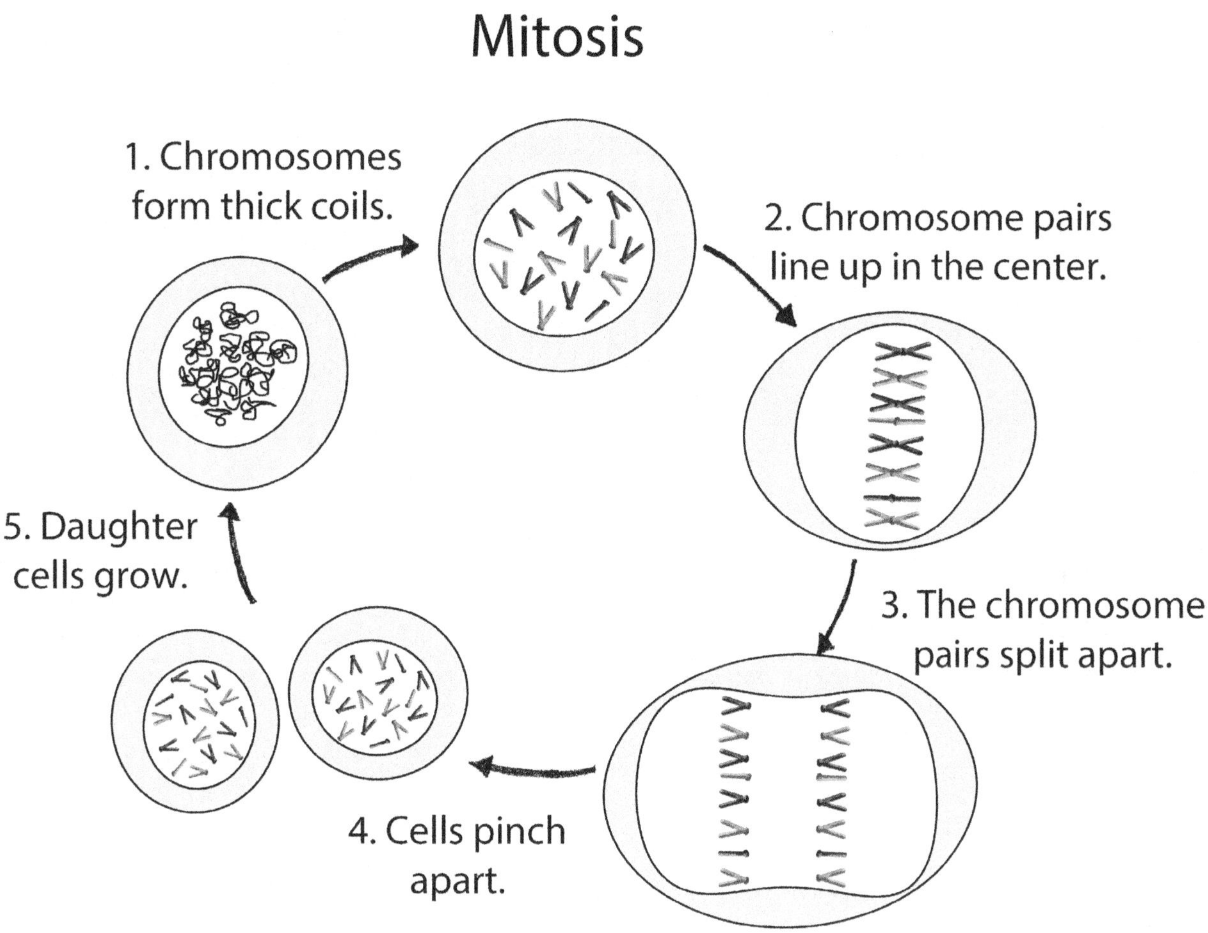

(Optional) Vertebrates Test

The students can complete the optional unit test found on pp. 148-151 in the Appendix. The answers for this test are found on pp. 80-81. A printable copy of the test can be downloaded for free from:

💻 https://elementalscience.com/blogs/resources/sq100

Vertebrates Test Answers

Vocabulary Matching (2 points each)

1. F
2. A
3. E
4. B
5. P
6. N
7. D
8. K
9. L
10. C
11. J
12. G
13. I
14. H
15. O
16. M

True and False (2 points each)

1. True
2. False (A fish uses its lateral line to detect movement.)
3. False (All birds lay eggs.)
4. True
5. False (The word amphibian comes from the Greek words "amphi" and "bios", meaning double life.)
6. True
7. True
8. True
9. True
10. False (Mammals generally take the longest to reproduce.)

Short Answer

1. (6 points) A fish breathes by drawing water into its mouth and passing that water through the

gills. In the gills, dissolved oxygen from the water is absorbed. Then water exits the fish's body through slits in the gills.

2. (3 points) Birds are efficient at flight due to their feathers, hollow bones, and strong breast muscles.
3. (8 points) First, embryos develop inside eggs. They hatch as either a tadpole or a larva. The tadpole or larva undergoes metamorphosis to develop into an adult. The adult matures, mates, and lays eggs so that the cycle can begin again.
4. (4 points) The four main groups of reptiles are crocodilians, tortoises and turtles, snakes, and lizards.
5. (8 points) In placental mammals, the female's placenta provides all the nutrients for the growing baby. They give birth to fully formed young. In contrast, marsupial mammals give birth to poorly developed young, which find their way into a pouch for the remainder of the development. Monotreme lay eggs, unlike any other mammals.
6. (15 points) Basic Classes of the Phyla Chordata and their basic characteristics:
   - Fish – Cold-blooded animals that live in water; they are covered with scales and breathe through gills.
   - Amphibians – Cold-blooded animals that live on land and in the water; they have soft skin.
   - Reptiles – Cold-blooded animals that lay eggs; they are covered with scales.
   - Birds – Warm-blooded animals that lay eggs; they have feathers and wings.
   - Mammals – Warm-blooded animals that feed their young milk; they are covered with fur.

Total points: 96

SciQuest

Milo's Mission Teacher Notes

Scientist Biography Unit

What you'll study in this unit-

Memory Work-

None

Lesson 11 Scientist Biography

Read: Gathering Information

Week's Readings

First, spark an interest by having the students read this week's story opener:

- Read Chapter 11 in *The PB Nanner Fluffer Test*: "Milo Visited the Library." (Your student will read about Milo meeting Addison at the library for their scientist project.)

> Note– There is no schedule for this lesson as it depends on the length of the book your students choose. We suggest you work with your students to spread the work across the week.

Step 1: Choose A Scientist

After the stduents read Milo's chapter, have them choose a scientist to study for the week. You can have the student choose a book from your local library or do a pre-prepared scientist biography unit for this week. We recommend choosing one of the Science Chunks units we offer, all of which also include hands-on science activites related to the scientist:

- https://elementalscience.com/collections/scientist-biography-units

The following information is from the student's lab journal:

> During the next few weeks, you are going to be researching and learning more about a scientist that has contributed to science. You need to begin your scientist biography project by choosing which scientist you will research. You can choose someone from the field of biology, like Galen, Robert Hooke, Carl Linnaeus, Gregor Mendel, Alexander Fleming, or Jacques Cousteau, or someone that has interested you.

Step 2: Research the Scientist

The following information is from the student's lab journal:

> Once you have chosen the scientist you would like to study, you can begin your research. Begin by looking for a biography on your chosen scientist at the library. Then, look for articles on the chemist in magazines, newspapers, encyclopedias, or on the Internet. You will need to know the following about your scientist to write your report:
>
> - Biographical information on the scientist (where he or she was born, his or her parents, siblings, and how he or she grew up)
> - The scientist's education (where he or she went to school, what kind of student he or she was, what he or she studied, and so on)
> - His or her scientific contributions (the research that he or she participated in, any significant discoveries he or she made, and the state of the world at the time of his or her contributions)

As you read over the material you have gathered, be sure to write down any facts you glean in your own words. You can do this on a separate sheet or on separate index cards - just make sure you have at least five pieces of information for each category. And don't forget to record a list of the resources you have used for your research.

Step 3: Create an Outline

The following information is from the student's lab journal:

Now that your research is completed, you are ready to begin the process of writing a report on your chosen scientist. You are going to organize the notes you took during step two into a formal outline which you will use next week to write the rough draft of your report. Use the outline template provided below as a guide. You should include information on why you chose the particular scientist in your introduction section. For the conclusion section of the outline, you need to include why you believe someone else should learn about your chosen scientist and your impression of the scientist (i.e. Did you like the scientist? Do you feel that they made a significant impact on their field of science?).

Step 4: Write a Rough Draft

The following information is from the student's lab journal:

Finally you are ready to take your outline and turn it into a rough draft. Simply take the points on your outline and combine or add in sentence openers and transitions to create a cohesive paragraph. Here's what your rough draft should look like:

- Paragraph 1 (from outline point I): introduce the scientist;
- Paragraph 2 (from outline point II): tell about the scientist's education;
- Paragraph 3-5 (from outline point III): share the scientist's contributions (one paragraph for each contribution);
- Paragraph 6 (from outline point IV): share your thoughts on the scientist and why someone should learn about him or her.

You can choose to hand write or type up your rough draft on a separate sheet of paper. However, keep in mind that you will need a typed version for step five.

Step 5: Revise to Create a Final Draft

The following information is from the student's lab journal:

Now that you have a typed, double-spaced rough draft, look over it one more time to make any changes you would like. Then, have your teacher or one of your peers look over the paper for you to correct any errors and bring clarity to any of the unclear sections. Once this is complete, make the necessary changes to your paper to create your final draft. Print out your final paper to share or use it to create a book or poster on the scientist.

You can have the students do an oral presention of their scientists if time allows.

SciQuest

Milo's Mission Teacher Notes

Planet Earth Unit

What you'll study in this unit-

Memory Work-

Lines on the Globe and World's Major Seas and Oceans (LJ p. 91)

Lesson 12 Grid Schedule

Supplies Needed	
Experiments	• Permanent marker, 1 Raw egg, 1 Hard-boiled egg
Optional Activities	• Modeling (or air dry) clay (red, orange, yellow, blue, and green), Ruler

Weekly Schedule

	Day 1	Day 2	Day 3	Day 4
Read	❑ Read Chapter 12: "Milo Shared a Cartoon" in *The PB Nanner Fluffer Test*	❑ Choose one of the encyclopedia readings to assign and discuss what was read using the questions provided	❑ *(Optional)* Choose another one of the encyclopedia readings to assign	❑ *(Optional)* Read one or both of the optional reading selections
Do	❑ Do the "Does the molten layer in Earth's core affect its movement?" experiment			❑ *(Optional)* Make a Model Planet Earth
Write	❑ Fill out the experiment sheet	❑ Write about what's inside the Earth	❑ Color and label the inside the Earth sketches ❑ Define mantle	❑ *(Optional)* Work on memorizing the Major Lines on the Globe

Lesson 12 List Schedule

Essentials

Read

- O Read Chapter 12: "Milo Shared a Cartoon" in *The PB Nanner Fluffer Test*.
- O Choose one (or more) of the encyclopedia readings to assign and discuss what was read using the questions provided.

Do

- O Do the "Does the molten layer in Earth's core affect its movement?" experiment. Fill out the lab report as you do the experiment.

Write

- O Write about what's inside the Earth.
- O Look up mantle. Add the definitions to the unit vocabulary sheet.
- O Label the inside the Earth sketches.

(Optional) Extras

Read

- O Read one or both of the optional reading selections.

Do

- O Make a Model Planet Earth.

Write

- O Work on memorizing the Major Lines on the Globe.

Supplies Needed	
Experiments	• Permanent marker, 1 Raw egg, 1 Hard-boiled egg
Optional Activities	• Modeling (or air dry) clay (red, orange, yellow, blue, and green), Ruler

Lesson 12 Science Focus: Inside The Earth

Read: Gathering Information

> **Key Idea**
> **The Earth consists of three main layers: the crust, the mantle, and the core.**

Week's Readings

First, spark an interest by having the students read this week's story opener:

- Read Chapter 12 in *The PB Nanner Fluffer Test*: "Milo Shared a Cartoon." (Your student will read about Milo and his cartoon, along with their co-op class project.)

Then follow the story up with some facts. You can choose which resource works best for your students:

- *Usborne Science Encyclopedia*: "Earth's Structure" pp. 180-181
- *Kingfisher Science Encyclopedia:* "Earth's Structure" pp. 8-9
- *Super Earth Encyclopedia:* "Layered Planet" pp. 14-15
- *Everything You Need to Ace Science in One Big Fat Notebook:* "The Earth's Structure and Composition" pp. 235-236

Discussion Questions

After the reading, use the following questions to guide your discussion time with the students.

1. What are the three main layers of the Earth? Be sure to include a detail or two about each. (The three main layers of the Earth are the crust, the mantle, and the core. The crust is the thinnest layer, which floats over the layers below. The mantle has a solid upper layer, which consists mostly of dense silicate rock. The core is extremely hot and consists of molten rock made from iron and nickel.)
2. How is the Earth's magnetic field created and why is it important? (The Earth's magnetic field is created by the electrical currents generated by the circulating molten iron in the core. The magnetic field protects the Earth from harmful radiation and particles from the sun.)
3. What causes seismic activity on the Earth? (Seismic activity on the Earth is caused by the heat escaping from the core.)
4. What have seismic scans of the Earth revealed about the mantle? (Seismic scans have revealed that there are probably two layers of circulating rock within the mantle.)

After the discussion time, have the students write a list of facts or a paragraph about what they read. Their writings should include at least a few facts from above.

(Optional) Reading Selections

At another point in the week, the students can check out these optional books:

- *The Wondrous Workings of Planet Earth: Understanding Our World and Its Ecosystems* by Rachel Ignotofsky
- *What's So Special about Planet Earth?* by Robert E. Wells

Do: Experimenting with Science

Experiment: Does the molten layer in Earth's core affect its movement?

Materials

You will need the following items:

- ✓ Permanent marker
- ✓ 1 Raw egg
- ✓ 1 Hard-boiled egg

Introduction

Read the following with your students, while they follow along in their lab journal.

Milo and his friends made a model of the Earth. The Earth consists of three main layers: the crust, the mantle, and the core. Some of these layers are solid, some of them are molten liquid. Each layer varies in thickness and consists of different elements. In this experiment, you are going to see how the composition of our Earth affect its motion.

Directions

Have the students do the following experiment testing how the liquid core of the Earth affects its movement under your supervision.

1. Mark the raw egg with a #1 and the hard-boiled egg with a #2.
2. Set the raw and the hard-boiled egg on the counter for twenty minutes to allow them to come up to room temperature.
3. Place both the eggs on the counter and try to spin each egg.
4. Observe how each one moves and record the time each egg spins on the chart.

Results and Explanation

When they are done, read the results and explanation to the students.

You should have found that the hard-boiled egg spins easily. It will proabably continue to spin for several second longer than the raw egg. The raw egg will also wobble more than the hard-boiled egg.

Discuss how the results from their experiment looked like or didn't look like the expected results. Then, read the following explanation to the students.

The material inside the egg's shell affected the spin of the egg's shell. The solid content spun with the shell because the center of gravity remains constant. The liquid content in the raw egg allows the center of gravity to shift, causing the movement of the spin to wobble and be a bit slower. The Earth's core is a mix of solid and liquid layers. This mix allows the Earth to spin and wobble as it rotates the sun.

(Optional) Hands-on Science Activities

At another point in the week, the students can do one or more of the following optional activities.

- **Model Earth –** You will need modeling (or air dry) clay (red, orange, yellow, blue, and green) and a ruler. Have the students begin by making a ball about ½ inch in diameter out of the yellow clay. (This represents the Earth's inner core.) Then, have the students make a flat circle out of the red clay and wrap it around the ball so that the ball is about 1 inch across. (This layer represents the Earth's outer core.) Next, have the students make a flat circle out of the orange clay and wrap it around the ball so that the ball is about 2 inches across. (This layer represents the Earth's mantle, the thickest layer.) Finally, have the students make a flat circle out of the blue clay and wrap it around the ball so that the ball is about 2 ½ inches across. Then, use the green clay to make a few flattened pieces to represent the continents, and layer them over your ball. (The blue and green clay layers represent the Earth's crust.) After the students have finished the ball, cut the ball in half and have them observe the layers.

Write: Keeping a Journal

Journaling Sheets

The students should have already recorded what they learned about what's inside the Earth on their notes page (LJ p. 99) and completed the experiment sheet (LJ pp. 96-97). On the notes page, the students can also answer the bonus question. Here is an answer for your reference as you discuss this with your students.

- **Bonus Question** – How would things change on Earth if the inside was all solid rock? (Answers will vary, but they should include something about how the tectonic plates wouldn't move, plus how it would affect the magnetic field around the globe and the geothermal features on the Earth's surface.)

Vocabulary

At some point in the week, have the students look up the following words and add the definitions to the unit vocabulary sheet on LJ pp. 92-94:

- **Mantle –** The mostly solid part of the Earth that lies between its crust and its core.

Sketch Assignment

See the next page for the correctly labeled sketch. (LJ p. 98)

(Optional) Memory Work

At another point in the week, the students can work on the following memory work - the Major Lines of Longitude and Latitude.

Inside the Earth

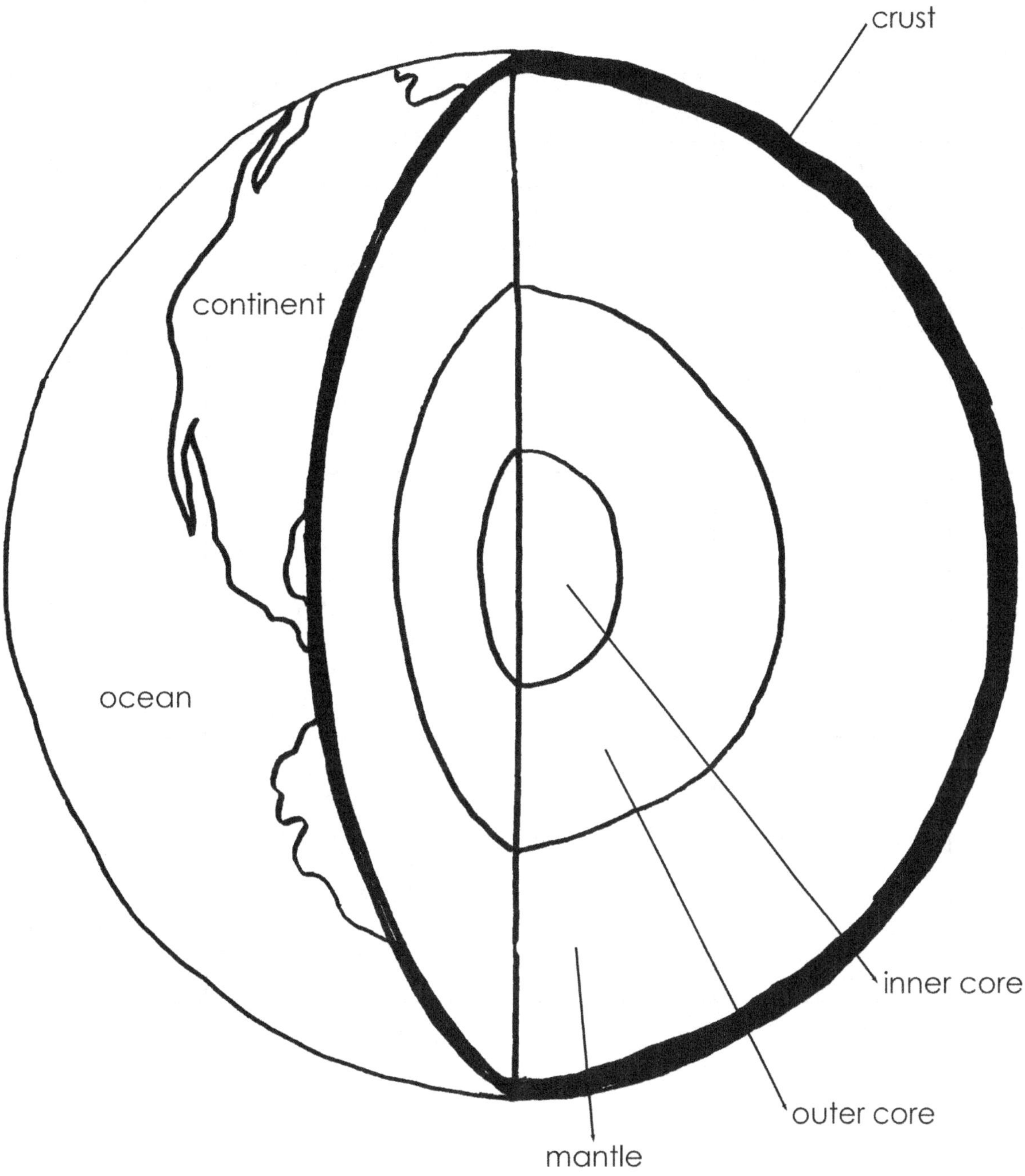

Lesson 13 Grid Schedule

Supplies Needed	
Experiments	• Blue balloon (with the continents drawn or printed on it), Flat map, Pin
Optional Activities	• Paper • Pumpkin, Paint

Weekly Schedule				
	Day 1	**Day 2**	**Day 3**	**Day 4**
Read	❑ Read Chapter 13: "Milo Solved the Riddle" in *The PB Nanner Fluffer Test*	❑ Choose one of the encyclopedia readings to assign and discuss what was read using the questions provided	❑ *(Optional)* Choose another one of the encyclopedia readings to assign	❑ *(Optional)* Choose another one of the encyclopedia readings to assign
Do	❑ Do the "Flat vs. Sphere" experiment		❑ *(Optional)* Create a map of your town	❑ *(Optional)* Make a Pumpkin Map
Write	❑ Fill out the experiment sheet	❑ Write about maps and mapping	❑ Color and label the lines on the globe sketch ❑ Define cartographer, lines of longitude, and lines of latitude	❑ *(Optional)* Work on memorizing the Major Lines on the Globe

Lesson 13 List Schedule

Essentials

Read

- ○ Read Chapter 13: "Milo Solved the Riddle" in *The PB Nanner Fluffer Test.*
- ○ Choose one of the encyclopedia readings to assign and discuss what was read using the questions provided.

Do

- ○ Do the "Flat vs. Sphere" experiment. Fill out the lab report as you do the experiment.

Write

- ○ Write about maps and mapping.
- ○ Look up cartographer, lines of longitude, and lines of latitude. Add the definitions to the unit vocabulary sheet.
- ○ Label the lines on the globe sketch.

(Optional) Extras

Read

- ○ Choose another one (or two) of the encyclopedia readings to assign.

Do

- ○ Create a map of your town.
- ○ Make a Pumpkin Map.

Write

- ○ Work on memorizing the Major Lines on the Globe.

Supplies Needed	
Experiments	• Blue balloon (with the continents drawn or printed on it), Flat map, Pin
Optional Activities	• Paper • Pumpkin, Paint

Lesson 13 Science Focus: Maps

Read: Gathering Information

Key Idea

Maps show physical information about an area.

Week's Readings

First, spark an interest by having the students read this week's story opener:

- Read Chapter 13 in *The PB Nanner Fluffer Test*: "Milo Solved the Riddle." (Your student will watch Milo and his friends as they travel around the co-op getting clues to solve a riddle.)

Then follow the story up with some facts. You can choose which resource works best for your students:

- *Kingfisher Science Encyclopedia*: "Maps and Mapping" pp. 46-47
- *Everything You Need to Ace Science in One Big Fat Notebook:* "Topographic Maps" p. 256

Discussion Questions

After all that reading, use the following questions to guide your discussion time with the students.

1. What is the purpose of a map? (The purpose of a map is to give geographical information about an area.)
2. What is the purpose of the lines that you find on a globe? (The purpose of the lines that can be found on a globe is to help us measure distances and find places on the globe.)
3. How are lines of latitude measured? (Lines of latitude are measured at angles north or south of the central latitude line, the equator.)
4. How do lines of longitude differ from lines of latitude? (Lines of longitude divide the globe in segments from pole to pole. Lines of longitude do not have an obvious reference point, like the equator, so one was arbitrarily chosen.)
5. How do cartographers represent a round Earth on a flat surface? (Cartographers represent a round Earth on flat surface by using projections.)
6. How are satellites used in mapping? (Satellites produce images of the Earth's surface that cartographers can use to produce more accurate maps of the Earth.)

After the discussion time, have the students write a list of facts or a paragraph about what they read. Their writings should include at least a few facts from above.

(Optional) Reading Selections

At another point in the week, the students can check out these optional books:

- *The Boy Who Loved Maps* by Kari Allen
- *Maps* by Aleksandra Mizielinska

Do: Experimenting with Science

Experiment: Flat vs. Sphere

Materials

You will need the following items:

- ✓ Blue balloon (with the continents drawn or printed on it)
- ✓ Flat map
- ✓ Pin

Introduction

Read the following with your students, while they follow along in their lab journal.

> Milo and his friends follow clues and maps to solve a riddle. Maps are graphical representations of the Earth. There are several different types of maps, such as world maps, relief maps, topographical maps, and street maps, but they all have one thing in common. All maps share information that helps the user to determine position, so it is very important that they are accurate. The Earth is a spherical object, but most maps are flat representations. In this experiment, you are going to examine if flat representations of the Earth are the same as spherical representations.

Directions

Have the students do the following experiment to compare different types of maps under your supervision.

1. Blow up your balloon so that it looks like a globe. (Note—If your balloon did not have the continents already printed on it, now is the time to draw them.)
2. Take your flat map and wrap it around the balloon, while attempting to match the continents up.
3. Record your observations.
4. Take the pin, pop your balloon.
5. Then, cut the balloon map between The Americas and Asia.
6. Try to stretch your balloon to match the map, and record your observations.

Results and Explanation

When they are done, read the results and explanation to the students.

> You should have found that it was very difficult, if not impossible, to perfectly match the flat map to the balloon globe without cutting the map. They should also see that it was very difficult, if not impossible, to perfectly match the popped balloon globe to the flat map.

Discuss how the results from their experiment looked like or didn't look like the expected results.

Then, read the following explanation to the students.

> It is virtually impossible to make a perfect representation of a spherical object on a flat surface. Either some of the object will be distorted or sections of the flat map will be missing. This is the reason that cartographers have come up with so many different projections or methods of representing the Earth on a flat surface.

(Optional) Hands-on Science Activities

At another point in the week, the students can do one or more of the following optional activities.

- **Create a Map** – Have the students create a street level map of your town, showing your home and various landmarks. Be sure they label the streets and landmarks. If you want to challenge them further, have them draw a grid with coordinate blocks and a scale for their map.
- **Pumpkin Map** – Have the students mark a pumpkin with the longitude and latitude lines. A pumpkin has natural longitude lines as well as a North and South Pole. Choose one of the longitude lines to be the Prime Meridian and then measure halfway in between the stem and bottom of the pumpkin to mark the Equator. Measure a third of the distance from the Equator and the stem and draw the Tropic of Cancer. Then, measure a quarter of the distance from the stem to the Equator and draw the Arctic Circle. Repeat for the bottom half of the pumpkin, labeling the Tropic of Capricorn and the Antarctic Circle. Then have the students add several more lines of longitude and latitude.

Write: Keeping a Journal

Journaling Sheets

The students should have already recorded what they learned about maps and mapping on their notes page (LJ p. 105) and completed the experiment sheet (LJ pp. 102-103). On the notes page, the students can also answer the bonus question. Here is an answer for your reference as you discuss this with your students.

- **Bonus Question** – What would you include on a map so that a stranger could find your favorite place in your town? (Answers will vary, but they should include important landmarks and roads.)

Vocabulary

At some point in the week, have the students look up the following words and add the definitions to the unit vocabulary sheet on LJ pp. 92-94:

- **Cartographer** – A person who makes maps.
- **Lines of Latitude (parallels)** – Lines that run parallel around the globe, dividing the globe into flat slices; they never meet.
- **Lines of Longitude (meridians)** – Lines that run from the North to the South pole, dividing the globe into segments; they all meet at the poles.

Sketch Assignment

Here is the correctly labeled sketch: (LJ p. 104)

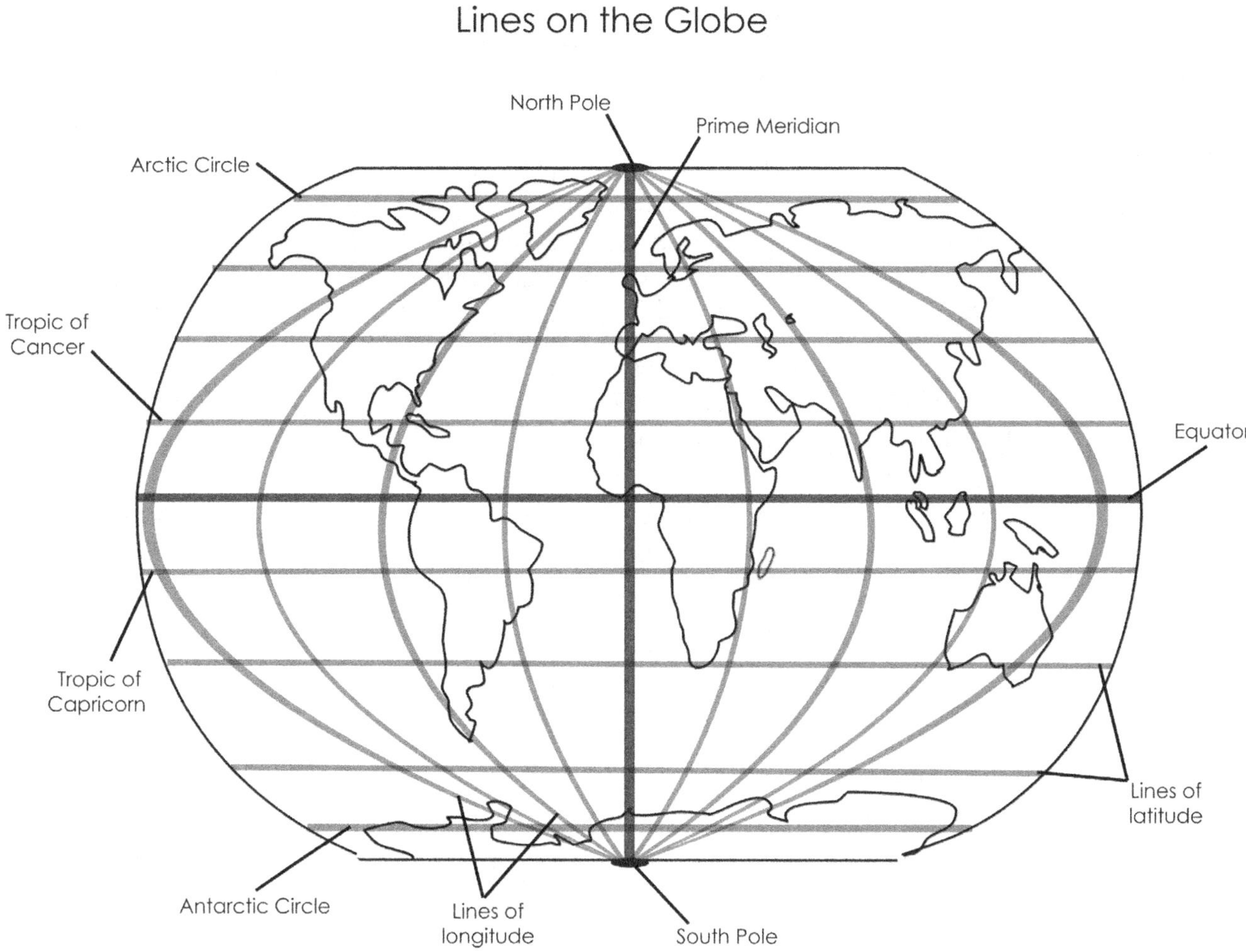

(Optional) Memory Work

At another point in the week, the students can work on the following memory work - the World's Major Oceans and Seas.

Lesson 14 Grid Schedule

Supplies Needed	
Experiments	• Pitcher for water, Water, Cookie sheet, Paper cup, Straw, Dirt or sand, Small rocks, Books, Tape
Optional Activities	• 2-Liter plastic bottle, Scissors, Tape, Paper, Soil, Water, Plastic tubing, Cup

Weekly Schedule

	Day 1	Day 2	Day 3	Day 4
Read	❑ Read Chapter 14: "Milo Caught a Ladybug" in *The PB Nanner Fluffer Test*	❑ Choose one of the encyclopedia readings to assign and discuss what was read using the questions provided	❑ *(Optional)* Choose another one of the encyclopedia readings to assign	❑ *(Optional)* Choose another one of the encyclopedia readings to assign
Do	❑ Do the "How is the course of a river determined?" experiment		❑ *(Optional)* Do the River Deposition activity	❑ *(Optional)* Explore a river online
Write	❑ Fill out the experiment sheet	❑ Write about rivers	❑ Color and label the river's course sketch ❑ Define delta, estuary, and source	❑ *(Optional)* Work on memorizing the World's Major Oceans and Seas

Lesson 14 List Schedule

Essentials

Read

- O Read Chapter 14: "Milo Caught a Ladybug" in *The PB Nanner Fluffer Test.*
- O Choose one of the encyclopedia readings to assign and discuss what was read using the questions provided.

Do

- O Do the "How is the course of a river determined?" experiment. Fill out the lab report as you do the experiment.

Write

- O Write about rivers.
- O Look up delta, estuary, and source. Add the definitions to the unit vocabulary sheet.
- O Label the river's course sketch.

(Optional) Extras

Read

- O Choose another one (or two) of the encyclopedia readings to assign.

Do

- O Do the River Deposition activity.
- O Explore a river online.

Write

- O Work on memorizing the World's Major Oceans and Seas.

Supplies Needed	
Experiments	• Pitcher for water, Water, Cookie sheet, Paper cup, Straw, Dirt or sand, Small rocks, Books, Tape
Optional Activities	• 2-Liter plastic bottle, Scissors, Tape, Paper, Soil, Water, Plastic tubing, Cup

Lesson 14 Science Focus: Rivers

Read: Gathering Information

> **Key Idea**
> **A river is a moving body of water. Streams can flow together to form a river.**

Week's Readings

First, spark an interest by having the students read this week's story opener:

- Read Chapter 14 in *The PB Nanner Fluffer Test*: "Milo Caught a Ladybug." (Your student will hang out with Milo and his class as they take a field trip to a local river.)

Then follow the story up with some facts. You can choose which resource works best for your students:

- *Usborne Science Encyclopedia*: "Rivers" pp. 190-191
- *Super Earth Encyclopedia:* "The Amazon River" pp. 124-125
- *Everything You Need to Ace Science in One Big Fat Notebook:* "Freshwater Ecosystems" p. 505

Discussion Questions

After all that reading, use the following questions to guide your discussion time with the students.

1. Where does the water in a river come from? (The water in a river comes from rainwater runoff, springs, or glacier flows.)
2. Describe a river's course. (A river begins in the mountains and flows downhill, where smaller streams or tributaries join it. As it grows and the land levels out, the river meanders through the landscape, forming large loops. Finally, it widens out into an estuary or delta before it reaches the sea.)
3. How do rivers erode the landscape around them? (Rivers erode by sweeping away any loose soil, sand, or rocks. As they sweep those away, they tear at the remaining rocks and chip away at the riverbed. The force of the river can also cause rocks to fracture and chemicals in the water can slowly dissolve the rocks as the river passes by.)
4. How is a load deposited by a river? (As a river slows down, it deposits the heavier material first. This is followed by the smaller, lighter particles, which creates layers.)
5. Where is a river delta found and what happens at this point in the river? (A river delta is found at the end of the river where it meets the sea. At this point in the river, fresh water mixes with salt water and the remaining sediment is deposited, creating narrower channels.)

After the discussion time, have the students write a list of facts or a paragraph about what they read. Their writings should include at least a few facts from above.

(Optional) Reading Selections

At another point in the week, the students can check out these optional books:

- *Great Rivers of the World* by Volker Mehnert

Do: Experimenting with Science

Experiment: How is the course of a river determined?

> Note-
> If your straw does not seal tightly, you may need to use some silicone sealant or air dry clay around the hole to prevent water from leaking out.

Materials

You will need the following items:

- ✓ Pitcher for water
- ✓ Water
- ✓ Cookie sheet
- ✓ Paper cup
- ✓ Straw
- ✓ Dirt or sand
- ✓ Small rocks
- ✓ Books
- ✓ Tape

Introduction

Read the following with your students, while they follow along in their lab journal.

> Milo and his friends went on a field trip to a river. A river is a natural flow of water from one area to a lake, sea, or ocean. Rivers typically contain fresh water and they are a very important part of the water cycle. Rivers are also home to a variety of wildlife and plant species. In this experiment, you are going to examine how a river's course is determined.

Directions

Have the students do the following experiment to see the course of a river under your supervision.

1. Poke a hole in the base of the cup, just large enough for the straw to fit tightly, using a pencil. Insert the straw and place the cup in the sink. Fill the cup halfway with water and watch what happens. If the water comes out of the straw like a stream, the cup is ready to use. If not, make the necessary adjustments.
2. Using several books, prop your cookie sheet up at an angle, so that water will flow down easily. Tape the cup to the top of the cookie sheet and fill it three quarters full with water. Observe what happens and record where the water went.
3. Dry the cookie sheet off and then sprinkle some dirt or sand all over the sheet. Refill the cup three quarters full with water, observe what happens and record where the water went.
4. Clean the cookie sheet off, sprinkle some dirt or sand all over the sheet, and place a few rocks in various places. Refill the cup three quarters full with water, observe what happens and record where the water went.

Results and Explanation

When they are done, read the results and explanation to the students.

> You should have seen that the plain stream (step 2) flowed straight down the sheet. The stream with the dirt (step 3) meandered a bit as it

made its way down the sheet. The stream with the dirt and rocks (step 4) meandered even more as it avoided the rocks in its way.

Discuss how the results from their experiment looked like or didn't look like the expected results. Then, read the following explanation to the students.

A river's course can be altered. As water makes its way down the mountain from the source, it encounters obstacles. The water can move some of these hindrances, but some of these impediments cannot be moved by the force of the water. Either way, these obstacles slow down and possibly alter the river's course. Other factors that contribute to how a river is formed are the amount of water and the angle of descent.

(Optional) Hands-on Science Activities

At another point in the week, the students can do one or more of the following optional activities.

- **River Depositation** – Have the students complete the activity suggested on p. 191 of the *Usborne Science Encyclopedia* to see how material carried by rivers is deposited in layers.
- **Explore a River** – Follow the directions from this website:
  http://tlc.howstuffworks.com/family/easy-outdoor-science-activities-for-kids1.htm

Write: Keeping a Journal

Journaling Sheets

The students should have already recorded what they learned about rivers on their notes page (LJ p. 111) and completed the experiment sheet (LJ pp. 108-109). On the notes page, the students can also answer the bonus question. Here is an answer for your reference as you discuss this with your students.

- **Bonus Question** – If a river were to disappear, what would happen to the surrounding environment? (Answers will vary.)

Vocabulary

At some point in the week, have the students look up the following words and add the definitions to the unit vocabulary sheet on LJ pp. 92-94:

- **Delta** – A fan-shaped system of streams that is created when a river splits into many smaller branches before it enters the sea.
- **Estuary** – A wide channel that forms where a river joins the sea.
- **Source** – The beginning of a river.

Sketch

See the next page for the correctly labeled sketch. (LJ p. 110)

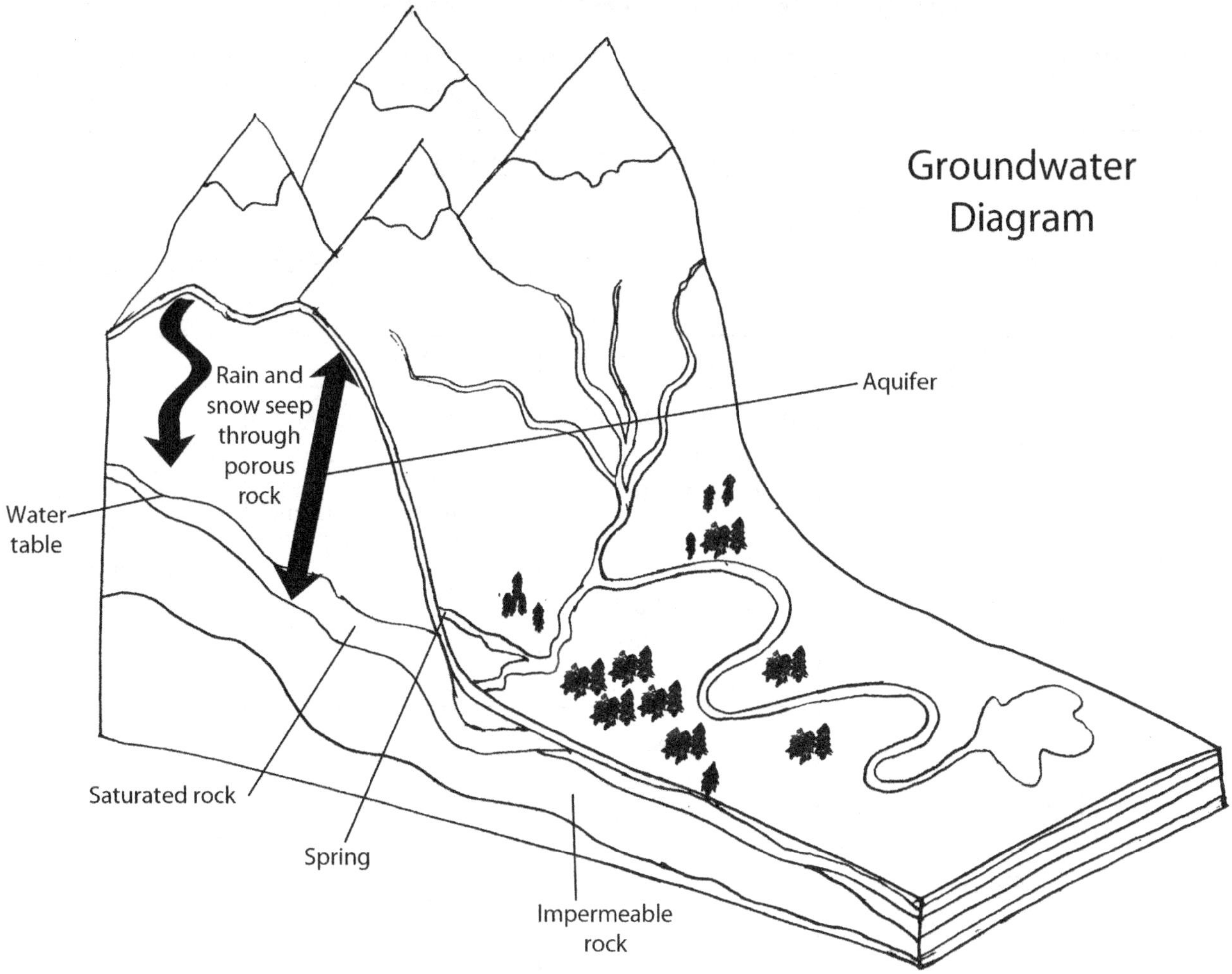

(Optional) Memory Work

At another point in the week, the students can work on the following memory work - the World's Major Oceans and Seas.

Lesson 15 Grid Schedule

Supplies Needed	
Experiments	• Aluminum bread pan or Plastic bin, Air dry clay, Water, Sand (1 cup), 2 Straws
Optional Activities	• Posterboard, Markers

Weekly Schedule				
	Day 1	**Day 2**	**Day 3**	**Day 4**
Read	❑ Read Chapter 15: "Milo Wallowed Alone" in *The PB Nanner Fluffer Test*	❑ Choose one of the encyclopedia readings to assign and discuss what was read using the questions provided	❑ *(Optional)* Choose another one of the encyclopedia readings to assign	❑ *(Optional)* Choose another one of the encyclopedia readings to assign
Do	❑ Do the "Surface Versus Deep Current" experiment		❑ *(Optional)* Explore a Tide Pool	❑ *(Optional)* Make a poster of the Ocean Zones
Write	❑ Fill out the experiment sheet	❑ Write about oceans ❑ Define oceanic ridge and oceanic trench	❑ Color and label the world's major seas and oceans sketch ❑ Define coast and tides	❑ *(Optional)* Work on memorizing the World's Major Oceans and Seas

Lesson 15 List Schedule

Essentials

Read

- O Read Chapter 15: "Milo Wallowed Alone" in *The PB Nanner Fluffer Test.*
- O Choose one of the encyclopedia readings to assign and discuss what was read using the questions provided.

Do

- O Do the "Surface Versus Deep Current" experiment. Fill out the lab report as you do the experiment.

Write

- O Write about oceans.
- O Look up coast, oceanic ridge, oceanic trench, and tides. Add the definitions to the unit vocabulary sheet.
- O Label the world's major seas and oceans sketch.

(Optional) Extras

Read

- O Choose another one (or two) of the encyclopedia readings to assign.

Do

- O Explore a Tide Pool.
- O Make a poster of the Ocean Zones.

Write

- O Work on memorizing the World's Major Oceans and Seas.

Supplies Needed	
Experiments	• Aluminum bread pan or Plastic bin, Air dry clay, Water, Sand (1 cup), 2 Straws
Optional Activities	• Posterboard, Markers

Lesson 15 Science Focus: Oceans

> **Key Idea**
>
> **The five oceans are the Arctic, Atlantic, Indian, Southern, and Pacific.**

Read: Gathering Information

Week's Readings

First, spark an interest by having the students read this week's story opener:

- Read Chapter 15 in *The PB Nanner Fluffer Test*: "Milo Wallowed Alone." (Your student will read about how Milo deals with the fallout of LB's reveal to the class during the field trip.)

Then follow the story up with some facts. You can choose which resource works best for your students:

- *Usborne Science Encyclopedia*: "Seas and Oceans" pp. 188-189
- *Kingfisher Science Encyclopedia*: "The Oceans" pp. 12-13, "The Ocean Floor" pp. 14-15
- *Everything You Need to Ace Science in One Big Fat Notebook:* "Tides" pp. 205-206

Discussion Questions

After all that reading, use the following questions to guide your discussion time with the students.

1. How much of the Earth's surface is covered with water? (Over two thirds of the Earth's surface is covered with water.)
2. How do currents work in the ocean? (In the ocean, as warm water around the tropics moves northward, it cools and sinks. The cooler water circulates back towards the south, creating a circular pattern.)
3. Why are the ocean depths so difficult to explore? (The ocean depths are difficult to explore because the underwater pressure is so great.)
4. Where do waves form and why do they crash on the shore? (Waves are formed in the open sea and are pushed by the wind. When they reach the shore, the lower part of the motion is slowed, causing the top of the wave to crest and crash on the coast.)

After the discussion time, have the students write a list of facts or a paragraph about what they read. Their writings should include at least a few facts from above.

(Optional) Reading Selections

At another point in the week, the students can check out these optional books:

- *Oceans: Geology and Weather (Science Readers)* by Yvonne Franklin
- *In One Tidepool: A Rhyming Marine Biology Book Perfect for the Classroom* by Anthony D Fredericks

Do: Experimenting with Science

Experiment: Surface Versus Deep Current

Materials

You will need the following items:

- ✓ Aluminum bread pan or Plastic bin
- ✓ Air dry clay
- ✓ Water
- ✓ Sand (1 cup)
- ✓ 2 Straws

Introduction

Read the following with your students, while they follow along in their lab journal.

> Addison told Milo a story about her trip to the ocean, which is constantly in motion due to several currents that affect it. These currents are caused by several factors, such as wind, density, and temperature. Surface currents are caused by the wind blowing over the surface of the ocean, which then forms waves. Deep water currents are caused by cold water sinking and warm water rising. Deep water currents can also cause underwater waves. In this experiment, you are going to examine how these two different currents can affect the ocean floor.

Directions

Have the students see two different types of ocean currents behave under your supervision.

1. Have your teacher cut a small hole at the center of one end of your pan that is 1½ inches from the bottom. Insert your straw into the hole, so that when you blow through it the air skips over the bottom of the pan. Use some of the air dry clay to hold the straw in place and prevent water from coming out of the hole.
2. Next, roll out two lengths of air dry clay that are the exact width of the pan and place them 2 inches from each end of the pan. (See diagram for a visual reference.)
3. Fill your pan two-thirds of the way with water and add ½ cup of sand to the end of the pan that has the straw. Make sure that the sand stays between the air dry clay roll and the end of the pan.
4. Make some deep water currents by blowing air through the straw at the end of the pan intermittently over one minute. Observe what happens and record your results.
5. Empty the pan, clean it out, refill it one third of the way with water and add ½ cup of sand in the same way you did in step # 4.
6. Make some surface currents by blowing air through the straw onto the surface of the water in the pan intermittently over one minute. Observe what happens and record your results.

Results and Explanation

When they are done, read the results and explanation to the students.

> You should have seen that both times the sand moved past the first barrier and possibly the second one as well. However, the students may see that when they create surface currents, the sand barely moves. This is because the water was too deep for the surface currents to be able to affect the bottom.

Discuss how the results from their experiment looked like or didn't look like the expected results. Then, read the following explanation to the students.

> Surface currents are caused by the wind blowing over the surface of the ocean, which then forms waves. They affect the ocean floor, but mostly when they are close to the shore, as the waves they create don't reach great depths. Deep water currents are caused by cold water sinking while warm water is rising. Deep water currents can also cause underwater waves, which can affect the ocean floor at great depths. In other words, both surface and deep water currents can move the ocean floor, but their effects are felt at different depths of the ocean.

(Optional) Hands-on Science Activities

At another point in the week, the students can do one or more of the following optional activities.

- **Ocean Zones** – Have the students make a poster depicting the different zones of the ocean. Have them include the depth ranges, a brief description of the conditions in the zones (i.e., temperature and pressure), and the typical animals found in that zone.
- **Tide Pools** – If you live near tide pools, take a field trip to explore them this week. If you do not, then visit the following video:

  https://www.youtube.com/watch?v=wpySMliPqNc

Write: Keeping a Journal

Journaling Sheets

The students should have already recorded what they learned about oceans on their notes page (LJ p. 117) and completed the experiment sheet (LJ pp. 114-115). On the notes page, the students can also answer the bonus question. Here is an answer for your reference as you discuss this with your students.

- **Bonus Question** – Would you like to live in the ocean? Why? (Answers will vary.)

Vocabulary

At some point in the week, have the students look up the following words and add the definitions to the unit vocabulary sheet on LJ pp. 92-94:

- **Coast** – The stretch of land that meets the sea.
- **Oceanic ridge** – A raised ridge on the seabed caused by the movement of the Earth's crustal plates.

- **Oceanic trench** – A deep trench in the seabed formed when one place of the Earth's crust moves under another.
- **Tides** – The daily movements of the sea up and down the shore; they are caused by the gravitational pull of the moon.

Sketch

Here is the correctly labeled sketch: (LJ p. 116)

The World's Major Seas & Oceans

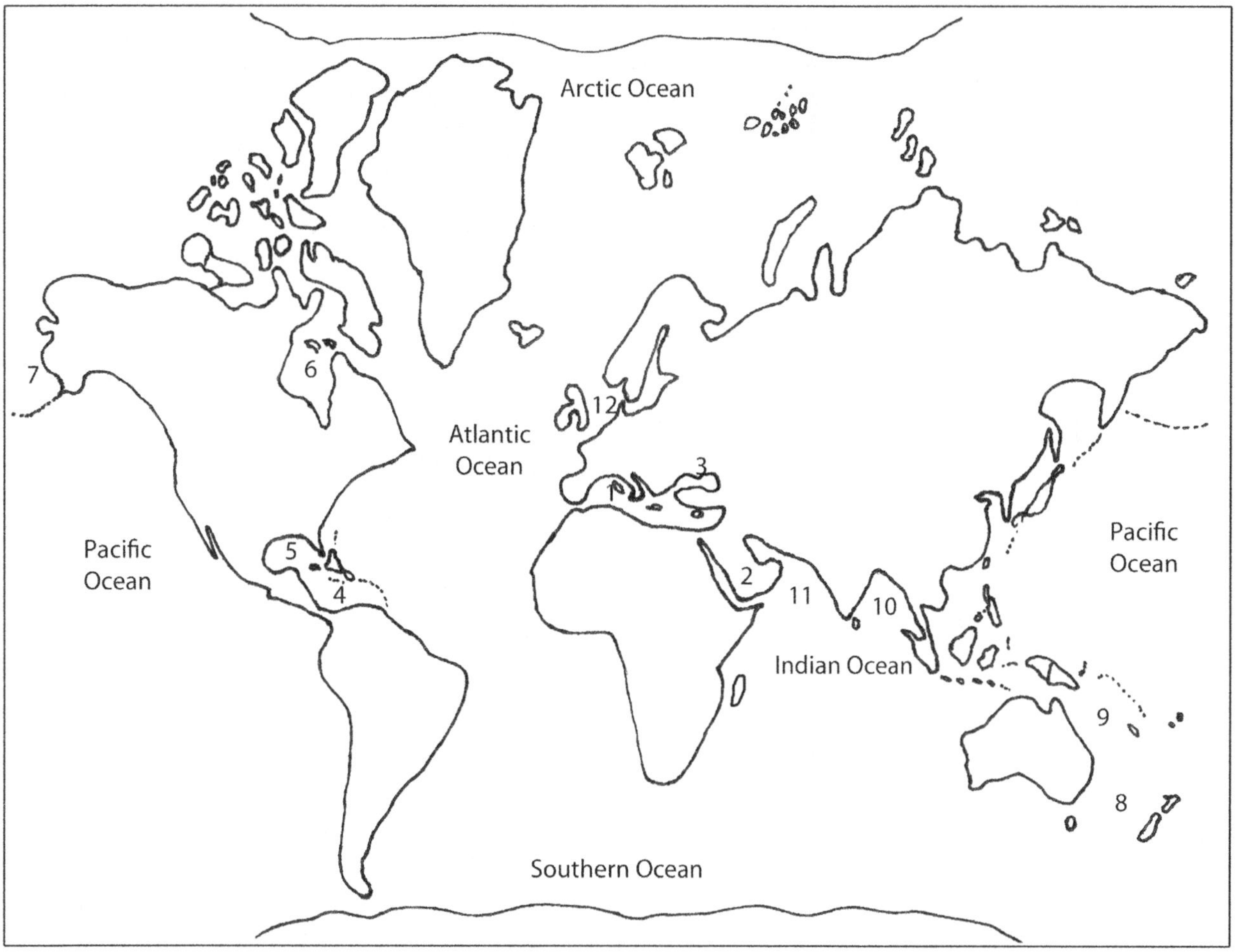

1. Mediterranean Sea	5. Gulf of Mexico	9. Coral Sea
2. Red Sea	6. Hudson Bay	10.Bay of Bengal
3. Black Sea	7. Bering Sea	11.Arabian Sea
4. Caribbean Sea	8. Tasman Sea	12.North Sea

(Optional) Memory Work

At another point in the week, the students can work on the following memory work - the World's Major Oceans and Seas.

Lesson 16 Grid Schedule

Supplies Needed	
Experiments	• Glacier Melt Model (You will need a cup, water, pebbles, and sand to make this.), Large cutting board with a handle, Large rubber band
Optional Activities	• Can of sweetened condensed milk, Cornstarch, Cereal box, Chocolate chips or nuts, Icing

Weekly Schedule				
	Day 1	**Day 2**	**Day 3**	**Day 4**
Read	❑ Read Chapter 16: "Milo Enjoyed Friendsgiving" in *The PB Nanner Fluffer Test*	❑ Choose one of the encyclopedia readings to assign and discuss what was read using the questions provided	❑ *(Optional)* Choose another one of the encyclopedia readings to assign	❑ *(Optional)* Choose another one of the encyclopedia readings to assign
Do	❑ Do the "What happens when glaciers melt?" experiment		❑ *(Optional)* Make Edible Glaciers	❑ *(Optional)* Research glaciers
Write	❑ Fill out the experiment sheet	❑ Write about glaciers	❑ Color and label the anatomy of a glacier sketch ❑ Define deposition, erratics, glacier, and moraine	❑ *(Optional)* Work on memorizing the World's Major Oceans and Seas

Lesson 16 List Schedule

Essentials

Read

- ○ Read Chapter 16: "Milo Enjoyed Friendsgiving" in *The PB Nanner Fluffer Test.*
- ○ Choose one of the encyclopedia readings to assign and discuss what was read using the questions provided.

Do

- ○ Do the "What happens when glaciers melt?" experiment. Fill out the lab report as you do the experiment.

Write

- ○ Write about glaciers.
- ○ Look up deposition, erratics, glacier, and moraine. Add the definitions to the unit vocabulary sheet.
- ○ Label the anatomy of a glacier sketch.

(Optional) Extras

Read

- ○ Choose another one (or two) of the encyclopedia readings to assign.

Do

- ○ Make Edible Glaciers.
- ○ Research glaciers.

Write

- ○ Work on memorizing the World's Major Oceans and Seas.

Supplies Needed	
Experiments	• Glacier Melt Model (You will need a cup, water, pebbles, and sand to make this.), Large cutting board with a handle, Large rubber band
Optional Activities	• Can of sweetened condensed milk, Cornstarch, Cereal box, Chocolate chips or nuts, Icing

Lesson 16 Science Focus: Glaciers

Read: Gathering Information

Key Idea

Glaciers are basically rivers of ice.

Week's Readings

First, spark an interest by having the students read this week's story opener:

- Read Chapter 16 in *The PB Nanner Fluffer Test*: "Milo Enjoyed Friendsgiving." (Your student will read about Milo and his friends celebrating Thanksgiving together.)

Then follow the story up with some facts. You can choose which resource works best for your students:

- *Kingfisher Science Encyclopedia*: "Glaciers and Ice Sheets" pp. 34-35
- *Super Earth Encyclopedia:* "Ice Caves: Mendenhall Glacier" pp. 106-107
- *Everything You Need to Ace Science in One Big Fat Notebook:* "Erosion and Deposition" pp. 254-256

Discussion Questions

After all that reading, use the following questions to guide your discussion time with the students.

1. Where are glaciers found? (Glaciers can be found on the highest mountains and near the poles.)
2. How do glaciers form? (Glaciers begin in high, cold places, such as mountains. Layers of fallen snow compact into hard ice. As more snow falls on top of the ice, the ice mass gets heavier and begins to move down the mountain. As it moves down the mountain, it slowly melts.)
3. What are some clues that let you know a glacier was once present in a valley? (You can know a valley once had a glacier by its deep, rounded, U-shape and by the moraine hills and other debris markers found in it.)

After the discussion time, have the students write a list of facts or a paragraph about what they read. Their writings should include at least a few facts from above.

(Optional) Reading Selections

At another point in the week, the students can check out these optional books:

- *Icebergs and Glaciers* by Seymour Simon
- *Glaciers (Eye To Eye With Endangered Habitats)* by Precious Mckenzie

Do: Experimenting with Science

Experiment: What happens when glaciers melt?

Materials

You will need the following items:

- ✓ Glacier Melt Model (Prepared the day before the experiment using cup, water, pebbles, and sand.)
- ✓ Large cutting board with a handle
- ✓ Large rubber band

Introduction

Read the following with your students, while they follow along in their lab journal.

> Milo's Thanksgiving spread ended up looking like a morraine field left behind a glacier. Glaciers are huge masses of ice that flow downhill. Glaciers are solid, powerful forces that can cut through the landscape, picking up rocks and soil along the way. In this experiment, you are going to look at what happens when those glaciers begin to melt.

Directions

Have the students make their own glacier and see what happens when it melts under your supervision.

Day Before the Experiment: Make the Model

1. Start making your Glacier Melt Model by pouring a layer of pebbles and then a layer of sand into the bottom of a cup.
2. Then, cover with water and place in the freezer.
3. Once the water has frozen, repeat the process again until the cup has been filled, giving you at least three layers.

Day of the Experiment

4. Prop your cutting board up in the sink, handle-side up, so that the board is at about a 45 degree angle.
5. Remove the Glacier Melt Model from the cup using a little warm water. Use the rubber band to secure it with the pebble/sand side down to the top of the cutting board.
6. Observe what happens as the ice melts.
7. Record your observations and the time it takes for the model to melt.

Results and Explanation

When they are done, read the results and explanation to the students.

> You should have seen that as the model melts, the pebbles and sand will slide down the board. Some will slide off in clumps and others will slide off in streams that form strange patterns.

Discuss how the results from their experiment looked like or didn't look like the expected results. Then, read the following explanation to the students.

> As a glacier melts, it affect the terrain. When a glacier melts and moves, it picks up rocks and dirt along the way. Then, as it melts, the glacier deposits the matter further down the hill. The valley just below a glacier will usually have a thick deposit of debris, called moraine, on the basin floor.

(Optional) Hands-on Science Activities

At another point in the week, the students can do one or more of the following optional activities.

- **Edible Glacier** – Make an edible glacier with the students. You can find the directions for this project here:

  https://elementalscience.com/blogs/science-activities/arctic-science-activities

- **Research Project** – Have the students research the changes in glaciers over the past twenty years. When they are done, have them write a one to three paragraph report about their findings.

Write: Keeping a Journal

Journaling Sheets

The students should have already recorded what they learned about glaciers on their notes page (LJ p. 123) and completed the experiment sheet (LJ pp. 120-121). On the notes page, the students can also answer the bonus question. Here is an answer for your reference as you discuss this with your students.

- **Bonus Question** – Why are glaciers blue? (Glaciers appear to be blue because of the way light is scattered as it passes through the ice. The longer, red wavelengths that are a part of white light are absorbed by the ice. The shorter, blue wavelengths that are part of white light are transmitted and scattered by the ice. In other words, the longer the path that light travels in the ice, the bluer the ice appears.)

  Note – This is a difficult bonus question, so if your students don't know the answer feel free to explain it to them.

Vocabulary

At some point in the week, have the students look up the following words and add the definitions to the unit vocabulary sheet on LJ pp. 92-94:

- **Deposition** – The dropping or leaving behind of rock and other debris by a glacier.
- **Erratics** – Large boulders that have been deposited away from their source by a glacier.
- **Glacier** – A mass of ice that gathers at the top of a land mass and slowly flows downhill.
- **Moraine** – Rock, clay, sand, and other debris left by a glacier at the valley floor.

Sketch

Here is the correctly labeled sketch: (LJ p. 122)

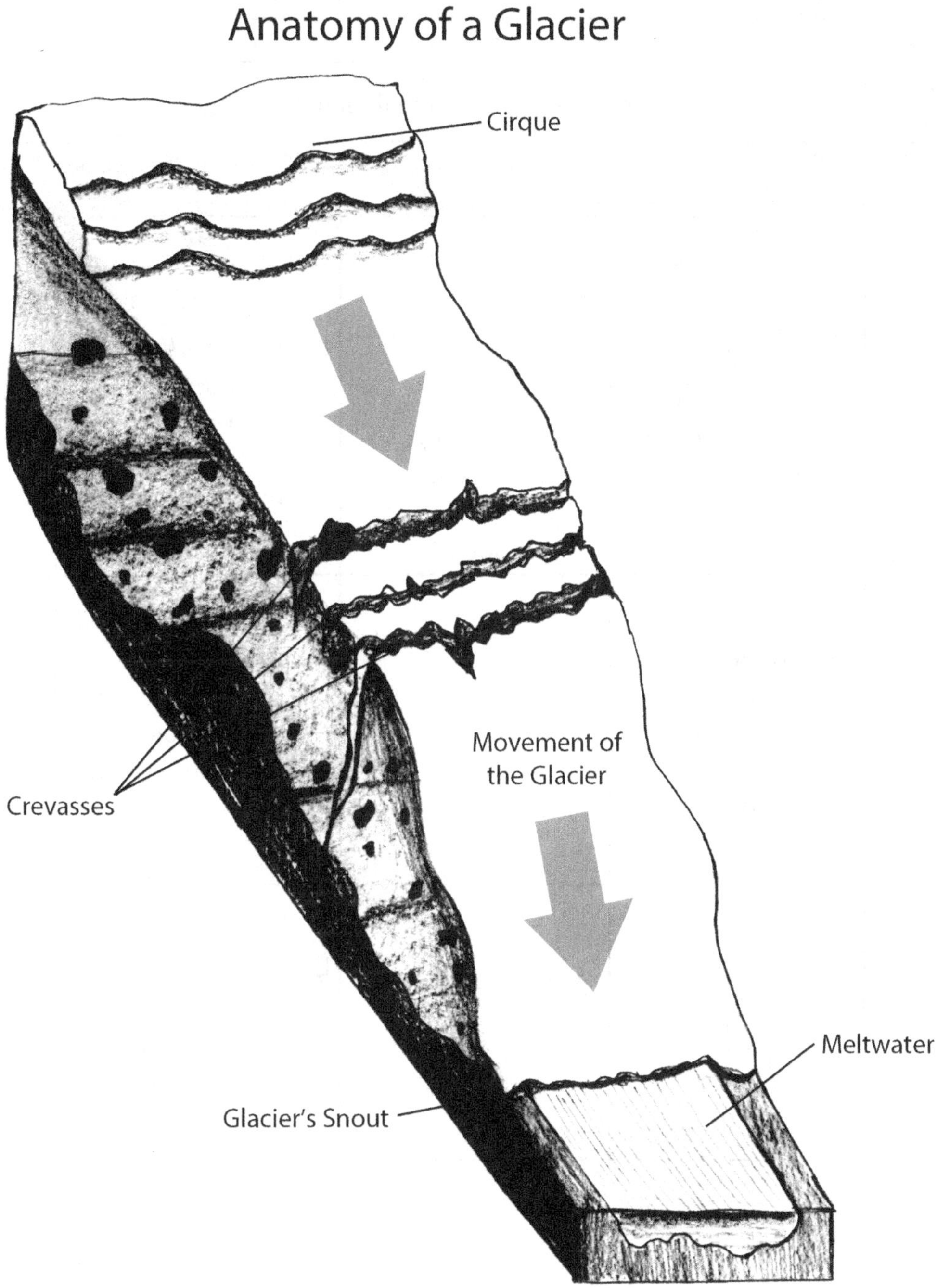

(Optional) Memory Work

At another point in the week, the students can work on the following memory work - the World's Major Oceans and Seas.

Lesson 17 Grid Schedule

Supplies Needed	
Experiments	• Paper towels, Cookie sheet, Wax paper, Rubber band, Water
Optional Activities	• Shoebox, Different plant and animal pictures or figurines

Weekly Schedule				
	Day 1	**Day 2**	**Day 3**	**Day 4**
Read	❑ Read Chapter 17: "Milo Executed the Plan" in *The PB Nanner Fluffer Test*	❑ Choose one of the encyclopedia readings to assign and discuss what was read using the questions provided	❑ *(Optional)* Choose another one of the encyclopedia readings to assign	❑ *(Optional)* Choose another one of the encyclopedia readings to assign
Do	❑ Do the "How do desert plants live with so little water?" experiment		❑ *(Optional)* Start working on a Habitat Diorama	❑ *(Optional)* Finish the Habitat Diorama
Write	❑ Fill out the experiment sheet	❑ Write about biomes and habitats	❑ Color and label a biome of the world sketch ❑ Define biome	❑ *(Optional)* Work on memorizing the World's Major Oceans and Seas

Lesson 17 List Schedule

Essentials

Read

- ○ Read Chapter 17: "Milo Executed the Plan" in *The PB Nanner Fluffer Test.*
- ○ Choose one of the encyclopedia readings to assign and discuss what was read using the questions provided.

Do

- ○ Do the "How do desert plants live with so little water?" experiment. Fill out the lab report as you do the experiment.

Write

- ○ Write about biomes and habitats.
- ○ Look up biome. Add the definitions to the unit vocabulary sheet.
- ○ Label and color a biome of the world sketch.

(Optional) Extras

Read

- ○ Choose another one (or two) of the encyclopedia readings to assign.

Do

- ○ Make a Habitat Diorama.

Write

- ○ Work on memorizing the World's Major Oceans and Seas.

Supplies Needed	
Experiments	• Paper towels, Cookie sheet, Wax paper, Rubber band, Water
Optional Activities	• Shoebox, Different plant and animal pictures or figurines

Lesson 17 Science Focus: Biomes

Key Idea

A biome is a larger region with a distinct climate.

Read: Gathering Information

Week's Readings

First, spark an interest by having the students read this week's story opener:

- Read Chapter 17 in *The PB Nanner Fluffer Test*: "Milo Executed the Plan." (Your student will read about Milo and his friends changing their plan for revenge.)

Then follow the story up with some facts. You can choose which resource works best for your students:

- *Usborne Science Encyclopedia*: "Plants and People" pp. 290-291
- *Kingfisher Science Encyclopedia*: "Biomes and Habitats" pp. 68-69
- *Super Earth Encyclopedia:* "Living Earth" pp. 188-203
- *Everything You Need to Ace Science in One Big Fat Notebook:* "Biomes" pp. 499-504

Discussion Questions

After all that reading, use the following questions to guide your discussion time with the students.

1. What determines the boundaries between biomes? (The boundaries of the different biomes are determined by variations in climate.)
2. What is the difference between a habitat and a biome? (A biome is a larger region with a distinct climate, which includes several smaller habitats. A habitat is an area where a given organism lives.)
3. What is a community? (A community is a group of plants and animals that live together in mutually beneficial relationships.)

After the discussion time, have the students write a list of facts or a paragraph about what they read. Their writings should include at least a few facts from above.

(Optional) Reading Selections

At another point in the week, the students can check out these optional books:

- *Habitats of the World* by DK
- *Understanding Biomes (Let's Explore Science)* by Jeanne Sturm

Do: Experimenting with Science

Experiment: How do desert plants live with so little water?

Materials

You will need the following items:

- ✓ Paper towels
- ✓ Cookie sheet
- ✓ Wax paper
- ✓ Rubber band
- ✓ Water

Introduction

Read the following with your students, while they follow along in their lab journal.

> Milo and his friends learned about habitats and biomes in co-op class. One of those biomes is the desert. The desert is a harsh environment in which only the hardiest plants and animals can survive. It is extremely hot during the day and very cold at night. The desert also receives very little rainfall. The plants and animals living there are well adapted to these conditions, like the camel, which can go for months without drinking water because its digestive system is extremely effective at extracting water. In this experiment, you are going to test how desert plants have adapted to survive with so little water.

Directions

Have the students use bits of paper towels to learn more about how desert plants survive under your supervision.

1. Moisten three paper towels with water until they are soaked, but not dripping, with water.
2. Lay one of the paper towels flat on the cookie sheet.
3. Roll the second paper towel up, secure it with the paper clips and lay it next to the flat paper towel on the cookie sheet.
4. Place the third paper towel flat on a piece of wax paper that is slightly larger, roll the two up together and secure with paper clips.
5. Then, lay the roll next to the other paper towels on the cookie sheet and set the sheet in a place where it will not be disturbed.
6. Check your paper towels the next day, make observations, and record the results.

Results and Explanation

When they are done, read the results and explanation to the students.

> You should have seen that the flat paper towel was completely dry and that the rolled paper towel was almost completely dry. The rolled paper towel plus the wax paper should still have been damp.

Discuss how the results from their experiment looked like or didn't look like the expected results. Then, read the following explanation to the students.

> Desert plants conserve water. The wax paper in the experiment simulated the waxy coating that most desert plants have. The rolled paper towel in the experiment represented the reduced number of stomata that are also found in a desert plant. Stomata are small pores in the leaves of plants that take in carbon dioxide and release water vapor. The cactus, a typical desert plant, has a thick waxy coating and a reduced number of stomata, both of which help it to live in an environment with very little water.

(Optional) Hands-on Science Activities

At another point in the week, the students can do the following optional activities.

- **Habitat Diorama** – Have the students choose to make a diorama of either the arctic or the deciduous forests. Then have them create that environment inside a shoebox and add the necessary plants and animals to it.

Write: Keeping a Journal

Journaling Sheets

The students should have already recorded what they learned about biomes and habitats on their notes page (LJ p. 129) and completed the experiment sheet (LJ p. 126-127). On the notes page, the students can also answer the bonus question. Here is an answer for your reference as you discuss this with your students.

- **Bonus Question** – Which biome would you like to live in and why? (Answers will vary)

Vocabulary

At some point in the week, have the students look up the following words and add the definitions to the unit vocabulary sheet on LJ pp. 92-94:

- **Biome** – A region of the Earth that contains unique plants and animals and is characterized by a distinct climate.

Sketch

See the next page for the correctly labeled sketches. (LJ p. 128)

(Optional) Memory Work

At another point in the week, the students can work on the following memory work - the World's Major Oceans and Seas.

Arctic

Desert

Grasslands

Rainforest

Taiga

Tundra

Lesson 18 Grid Schedule

Supplies Needed	
Experiments	• Cup, Water, Plastic Baggie, Rubber Band
Optional Activities	• Posterboard • 2-Liter bottle or a large glass jar, Soil, Several small plants

Weekly Schedule

	Day 1	Day 2	Day 3	Day 4
Read	❑ Read Chapter 18: "Milo Celebrated the Holidays" in *The PB Nanner Fluffer Test*	❑ Choose one of the encyclopedia readings to assign and discuss what was read using the questions provided	❑ *(Optional)* Choose another one of the encyclopedia readings to assign	❑ Read "Epilogue: Milo Has His Friends Over" in *The PB Nanner Fluffer Test*
Do	❑ Do the "Can I recreate the water cycle?" experiment		❑ *(Optional)* Make a Water Cycle Poster	❑ *(Optional)* Create your own Greenhouse
Write	❑ Fill out the experiment sheet	❑ Write about natural cycles ❑ Color and label the nitrogen cycle sketch	❑ Color and label the carbon cycle sketch ❑ Define natural cycle and greenhouse gas	❑ *(Optional)* Take the Unit 3 Test

Lesson 18 List Schedule

Essentials

Read

- ○ Read Chapter 18: "Milo Celebrated the Holidays" in *The PB Nanner Fluffer Test.*
- ○ Choose one of the encyclopedia readings to assign and discuss what was read using the questions provided.
- ○ Read "Epilogue: Milo Has His Friends Over" in *The PB Nanner Fluffer Test*

Do

- ○ Do the "Can I recreate the water cycle?" experiment. Fill out the lab report as you do the experiment.

Write

- ○ Write about natural cycles.
- ○ Look up natural cycle and greenhouse gas. Add the definitions to the unit vocabulary sheet.
- ○ Label the nitrogen and carbon cycle sketches.

(Optional) Extras

Read

- ○ Choose another one (or two) of the encyclopedia readings to assign.

Do

- ○ Make a Water Cycle Poster.
- ○ Create your own Greenhouse.

Write

- ○ Take the Unit 3 Test.

Supplies Needed	
Experiments	• Cup, Water, Plastic Baggie, Rubber Band
Optional Activities	• Posterboard • 2-Liter bottle or a large glass jar, Soil, Several small plants

Lesson 18 Science Focus: Natural Cycles

Read: Gathering Information

Key Idea

Natural cycles show how nutrients are recycled in nature.

Week's Readings

First, spark an interest by having the students read this week's story opener:

- Read Chapter 18 in *The PB Nanner Fluffer Test*: "Milo Celebrated the Holidays." (Your student will read about Milo and his family at the co-op's holiday showcase.)

Then follow the story up with some facts. You can choose which resource works best for your students:

- *Usborne Science Encyclopedia*: "Natural Cycles" pp. 292-293, "Nutrient Cycles" pp. 334-335
- *Usborne Illustrated Dictionary of Science:* "Carbon Cycle and Nitrogen Cycle" pp. 234-235
- *Everything You Need to Ace Science in One Big Fat Notebook:* "Water Cycle, Nitrogen Cycle, and Carbon Cycle" pp. 491-494

Discussion Questions

After all that reading, use the following questions to guide your discussion time with the students.

1. Briefly describe the nitrogen cycle. (Lightning forms nitrates in the air and bacteria form other nitrates in the soil. These nitrates are taken up by plants, which are then eaten by animals. When these plants and animals die, they release nitrogen back into the environment and the cycle begins again.)
2. Briefly describe the carbon cycle. (Plants take in carbon dioxide from the air and use it to make carbohydrates. Living things, like animals, ingest those carbohydrates and break them down for energy. This process produces carbon dioxide, which is released into the air and the cycle begins again.)
3. Briefly describe the water cycle. (Water falls as rain. It drains into the rivers and out to the sea. Then, the sun heats up the water, forming vapor, which rises to form clouds. Once the clouds become heavy with water, the cycle begins again.)
4. How are the natural cycles being upset? (The natural cycles are being upset by the presence of pollution and additional wastes and chemicals used by humans. This is creating more carbon dioxide in the air and more nitrogen in the soil than is needed.)

After the discussion time, have the students write a list of facts or a paragraph about what they read. Their writings should include at least a few facts from above.

(Optional) Reading Selections

At another point in the week, the students can check out these optional books:

- *A Drop Around the World: The Science Of Water Cycles On Planet Earth For Kids* by Barbara Shaw McKinney

📖 *It's Cycle Time! Carbon Cycle, Nitrogen Cycle and Water Cycle in an Ecosystem Explained* by Baby Professor

Do: Experimenting with Science

Experiment: Can I recreate the water cycle?

Materials

You will need the following items:

- ✓ Cup
- ✓ Water
- ✓ Plastic Baggie
- ✓ Rubber Band

Introduction

Read the following with your students, while they follow along in their lab journal.

Milo helped his sister recreate the water cycle, which shows how water changes forms on the Earth, on a cookie.

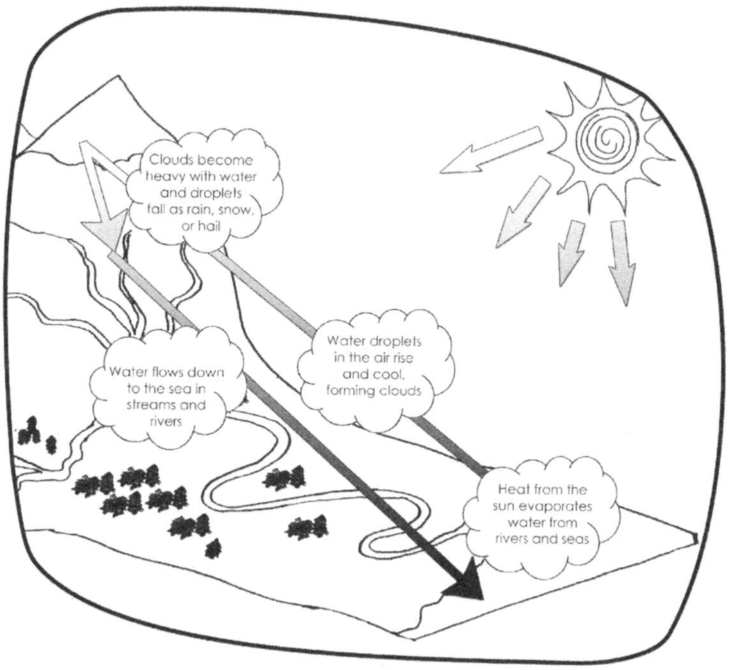

In today's experiment, you are going to try to create your very own mini-version of the water cycle.

Directions

Have the students use a ziploc baggie to see how the water cycle works under your supervision.

1. Pour a little warm water into the bottom of your cup.
2. Cover it with the plastic baggie and use the rubber band to secure it in place.
3. Set the cup in a place where it will receive direct sunlight for the next two hours. (Note—If

you do not have a sunny place to set your cup, set it under a desk lamp instead.)

4. Check the glass after 2 hours.
5. Observe and record what has happened.

Results and Explanation

When they are done, read the results and explanation to the students.

> You should have noticed that the plastic baggie has many droplets of water on it. They should also see that several of these droplets have come together to form bigger droplets, which then fall to the bottom of the cup.

Discuss how the results from their experiment looked like or didn't look like the expected results. Then, read the following explanation to the students.

> Water from the oceans and rivers is being evaporated by the Sun. The water vapors cool and come together to form clouds. When enough water vapor is in the clouds, they become heavy and rain begins to fall. The rain fills the rivers and oceans on the Earth once again.

(Optional) Hands-on Science Activities

At another point in the week, the students can do one or more of the following optional activities.

- **Water Cycle Poster** – Have the students make a poster detailing the water cycle.
- **Make a Greenhouse** – Use a 2-Liter bottle or a large glass jar to make a terrarium or mini greenhouse. Fill the bottom of your container with moist soil and plant several small plants in it. Cover and set it in the sun. You should see that the water follows through the water cycle as the plants grow over time.

Write: Keeping a Journal

Journaling Sheets

The students should have already recorded what they learned about natural cycles on their notes page (LJ p. 135) and completed the experiment sheet (LJ pp. 132-133). (Normally, I share a bonus question, but there isn't one this week.)

Vocabulary

At some point in the week, have the students look up the following words and add the definitions to the unit vocabulary sheet on LJ pp. 92-94:

- **Natural cycle** – The exchanging of essential elements, such as nitrogen, carbon, and oxygen.
- **Greenhouse gas** – Any gas that traps heat from the Sun, such as carbon dioxide.

Sketch

Here are the correctly labeled sketches: (LJ p. 134)

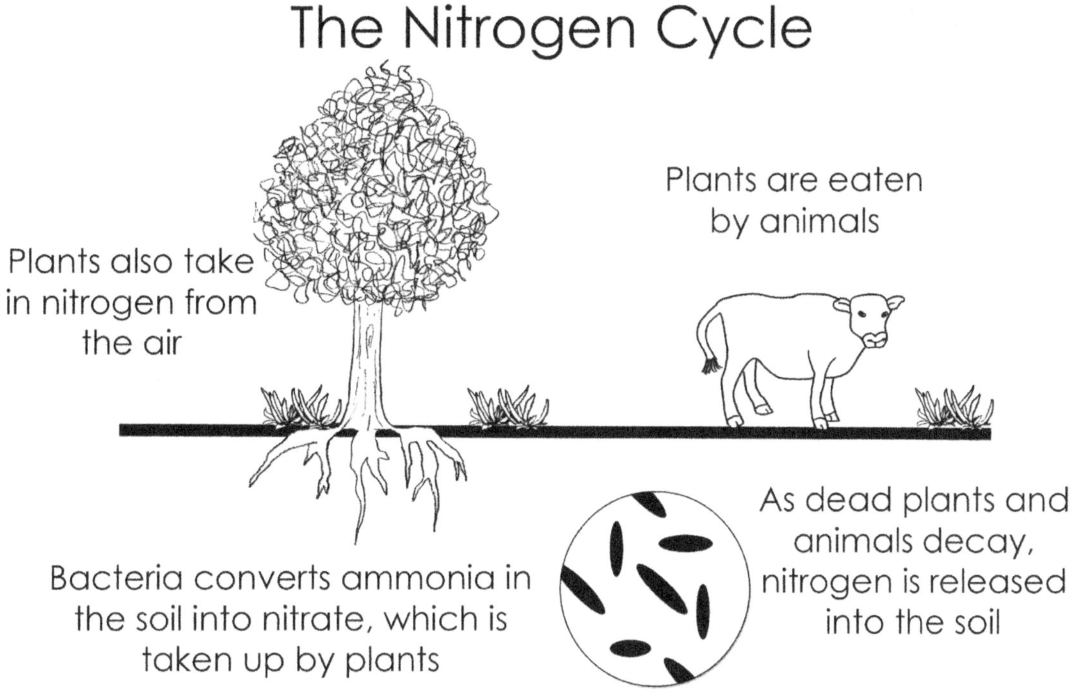

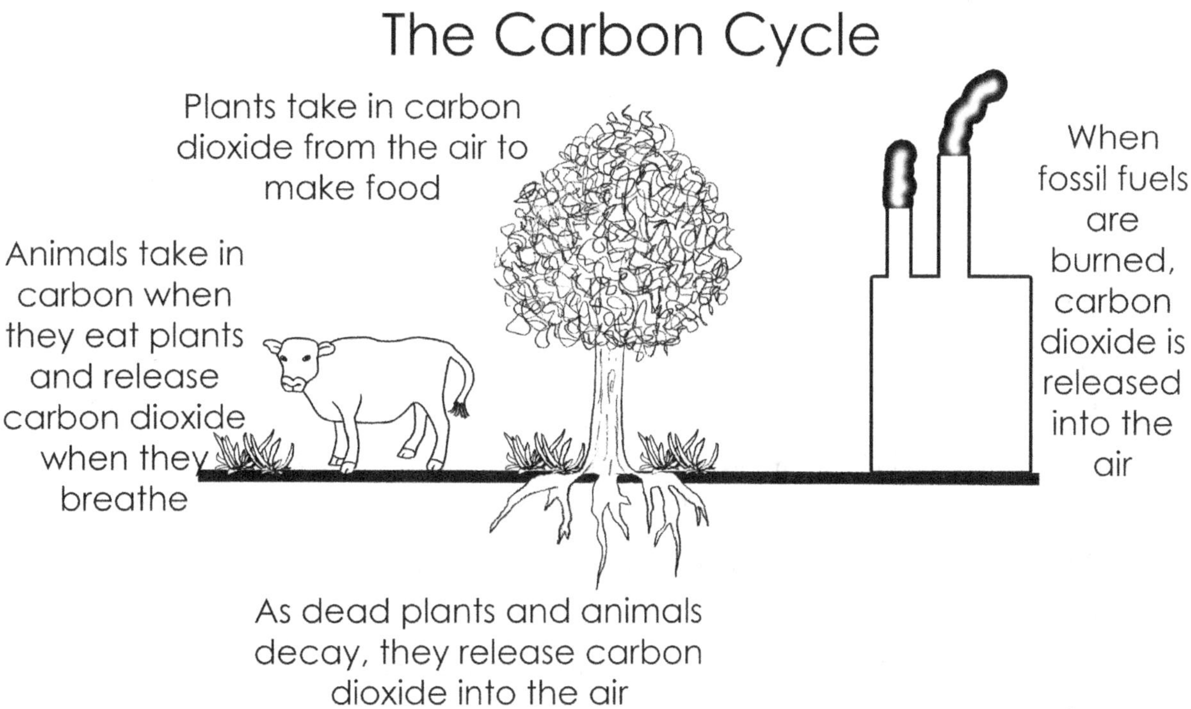

(Optional) Planet Earth Test

The students can complete the optional unit test found on pp. 152-156 in the Appendix. The answers for this test are found on pp. 130-131. A printable copy of the test can be downloaded for free from:

- https://elementalscience.com/blogs/resources/sq100

Planet Earth Test Answers

Vocabulary Matching (2 points each)

1. L
2. F
3. O
4. D
5. G
6. R
7. K
8. I
9. E
10. H
11. A
12. C
13. M
14. B
15. Q
16. N
17. J
18. P

Note -
If you need a digital copy of the unit test, you can download one for free from this program's resource page. Just head to:
https://elementalscience.com/blogs/resources/sq100

True and False (2 points each)

1. True
2. True
3. False (Satellites do help cartographers to produce more accurate maps.)
4. True
5. False (A river begins inland and flows into a lake, sea, or ocean.)
6. True
7. False (Over two-third's of the Earth's surface is covered with salt water.)
8. False (The gravitational pull of the Moon affects the oceans on the Earth, creating tides.)
9. False (Glaciers can also be found in the arctic regions, near the poles.)
10. True
11. True
12. False (Pollution has had an effect on the natural cycles.)

13. True
14. True

Short Answer

1. (6 points) The three main layers of the Earth are the crust, the mantle, and the core. The crust is the thinnest layer, which floats over the layers below. The mantle has a solid upper layer, which consists mostly of dense silicate rock. The core is extremely hot and consists of molten rock made from iron and nickel.
2. (5 points) Answers below:

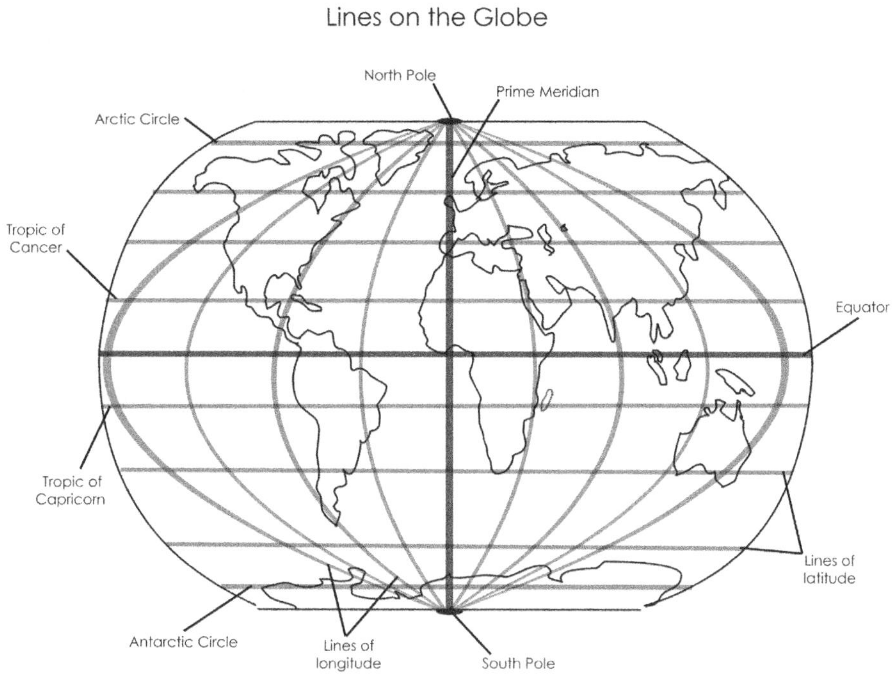

3. (5 points) A river begins in the mountains and flows downhill, where smaller streams or tributaries join it. As it grows and the land levels out, the river meanders through the landscape, forming large loops. Finally, the river widens out into an estuary or delta before it reaches the sea.
4. (4 points) In the ocean, as warm water around the tropics moves northward, it cools and sinks. The cooler water circulates back towards the south, creating circular pattern.
5. (5 points) Glaciers begin in high, cold places, such as mountains. Layers of fallen snow compact into hard ice. As more snow falls on top of the ice, the ice mass gets heavier and begins to move down the mountain. As it moves down the mountain, the ice slowly melts.
6. (6 points) Water falls as rain. It drains into the rivers and out to the sea. Then, the sun heats up the water, forming vapor, which rises to form clouds. Once the clouds become heavy with water, the cycle begins again.
7. (3 points) A biome is a larger region with a distinct climate, which includes several smaller habitats. A habitat is an area where a given organism lives.

Total points: 98

SciQuest

Appendix Templates

Microscope Worksheet

Sample Name

____

Description

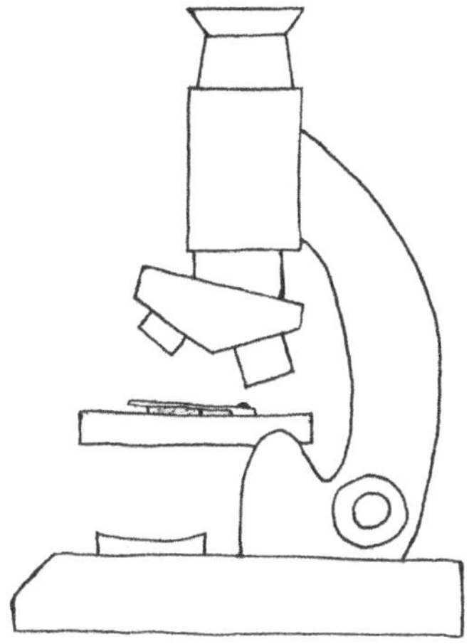

____
____
____
____
____
____
____
____
____
____
____

Sketch

Magnification Power ______x

Magnification Power ______x

SciQuest

GLOSSARY

Level 100 Glossary

A

- Amphibian – A cold-blooded vertebrate that lives partly in water and partly on land.
- Autotroph – A living thing that makes its own food.

B

- Biome – A region of the Earth that contains unique plants and animals and is characterized by a distinct climate.

C

- Carnivore – An animal with specially shaped teeth that feeds mainly on meat.
- Cartographer – A person who makes maps.
- Cell – A tiny unit of living matter, the basic building block of all life.
- Classification – A way of identifying or grouping living things.
- Clutch – A set of eggs.
- Coast – The stretch of land that meets the sea.
- Crop – A pouch in a bird's digestive system where swallowed food is stored.

D

- Delta – A fan-shaped system of streams that is created when a river splits into many smaller branches before it enters the sea.
- Deposition – The dropping or leaving behind of rock and other debris by a glacier.

E

- Ectotherm – An animal whose temperature varies with its surroundings; ectothermic animals are also called cold-blooded.
- Endotherm – An animal with a constant temperature; endothermic animals are also called warm-blooded.
- Erratics – Large boulders that have been deposited away from their source by a glacier.
- Estuary – A wide channel that forms where a river joins the sea.

F

- Fledgling – A young bird at the time it leaves the nest.

G

- Gills – An organ used to breathe underwater.
- Gizzard – A chamber in an animal's stomach that grinds up food.
- Glacier – A mass of ice that gathers at the top of a land mass and slowly flows downhill.
- Greenhouse gas – Any gas that traps heat from the Sun, such as carbon dioxide.

H

- Hemisphere – One of the halves of the Earth created by an imaginary division along the equator.
- Herbivore – An animal that eats only plant food.
- Heterotroph – A living thing that eats other living things.

I

J

K

- Kingdom – The highest category into which living things are classified.

L

- Lines of Latitude (parallels) – Lines that run parallel around the globe, dividing the globe into flat slices; they never meet.
- Lines of Longitude (meridians) – Lines that run from the North to the South pole, dividing the globe into segments; they all meet at the poles.

M

- Mammals – A warm-blooded animal with hair that feeds its young on milk.
- Mammary gland – The milk-producing organ of a female mammal.
- Mantle – The mostly solid part of the Earth that lies between its crust and its core.
- Migration – A journey by an animal to a new habitat.

- Mitosis – The division of a cell nucleus to produce two identical cells.
- Monotreme – A mammal that lays eggs, such as a duck-billed platypus.
- Moraine – Rock, clay, sand, and other debris left by a glacier at the valley floor.

N

- Natural cycle – The exchanging of essential elements, such as nitrogen, carbon, and oxygen.

O

- Oceanic ridge – A raised ridge on the seabed caused by the movement of the Earth's crustal plates.
- Oceanic trench – A deep trench in the seabed formed when one place of the Earth's crust moves under another.
- Omnivore – An animal that eats both plant and animal food.

P

Q

R

- Reptile – A cold-blooded vertebrate with scaly skin.

S

- Scales – Small, overlapping plates that protect the skin.
- Source – The beginning of a river.
- Species – A group of living things than can breed together in the wild.
- Swim bladder – A gas-filled bag that helps a fish to float in water.

T

- Tadpole – The immature form of a frog or toad.
- Tides – The daily movements of the sea up and down the shore; they are caused by the gravitational pull of the moon.

U

V

- Vertebrate – An animal with a backbone.

W

X

Y

Z

SciQuest

Unit Tests

ANIMAL BASICS UNIT TEST

VOCABULARY MATCHING

1. ___ Autotroph
2. ___ Camouflage
3. ___ Carnivore
4. ___ Cell
5. ___ Classification
6. ___ Hemisphere
7. ___ Herbivore
8. ___ Heterotroph
9. ___ Kingdom
10. ___ Migration
11. ___ Mitosis
12. ___ Omnivore
13. ___ Predator
14. ___ Prey
15. ___ Species

A. A tiny unit of living matter, the basic building block of all life.

B. The division of a cell nucleus to produce two identical cells.

C. A way of identifying or grouping living things.

D. The highest category into which living things are classified.

E. A group of living things than can breed together in the wild.

F. One of the halves of the Earth created by an imaginary division along the equator.

G. A journey by an animal to a new habitat.

H. A living thing that makes its own food.

I. A living thing that eats other living things.

J. An animal with specially shaped teeth that feeds mainly on meat.

K. An animal that eats only plant food.

L. An animal that eats both plant and animal food.

M. The way animals hide by blending in with their surroundings.

N. An animal that kills and eats other animals.

O. An animal that is killed and eaten by other animals.

TRUE AND FALSE

1. __________ Plant cells are identical to animal cells in all ways.

2. __________ Individual cells group together to form tissues, different types of tissues group together to form organs, organs group together to form systems, and the systems group together to form a living individual called an organism.

3. _________ Carolus Linnaeus devised a system for classifying living things.

4. _________ DNA fingerprinting does not help scientists with classifying living things.

5. _________ Animals migrate to breed and to find food.

6. _________ Amphibians are the most common migrants.

7. _________ Omnivores can eat almost anything edible, both plants and meat, from their environment.

8. _________ Birds have the same kind of stomanchs as humans do.

9. _________ An adaptation is a change that an animal or plant makes to better suit the environment it lives in.

10. _________ A defense is the way that an animal or plant protects itself.

Short Answer

1. What are the two stages of cell division in a plant cell? What happens in each of them?

2. What are the five kingdoms for classifications and their basic characteristics?

3. Name two ways that birds use to navigate.

4. How do a carnivore's teeth differ from an herbivore's teeth?

5. What are genotypic adaptations? Phenotypic adaptations?

Vertebrates Unit Test

Vocabulary Matching

1. Gills ___
2. Swim bladder ___
3. Vertebrate ___
4. Amphibian ___
5. Tadpole ___
6. Ectotherm ___
7. Reptile ___
8. Scales ___
9. Clutch ___
10. Crop ___
11. Gizzard ___
12. Fledgling ___
13. Endotherm ___
14. Mammals ___
15. Mammary gland ___
16. Monotreme ___

A. A set of eggs.

B. An animal whose temperature varies with its surroundings, also known as cold-blooded.

C. The milk-producing organ of a female mammal.

D. An organ used to breathe underwater.

E. A pouch in a bird's digestive system where swallowed food is stored.

F. A cold-blooded vertebrate that lives partly in water and partly on land.

G. A cold-blooded vertebrate with scaly skin.

H. A gas-filled bag that helps a fish to float in water.

I. Small, overlapping plates that protect the skin.

J. A mammal that lays eggs, such as a duck-billed platypus.

K. A chamber in an animal's stomach that grinds up food.

L. A warm-blooded animal with hair that feeds its young on milk.

M. An animal with a backbone.

N. A young bird at the time it leaves the nest.

O. The immature form of a frog or toad.

P. An animal with a constant temperature, also known as warm-blooded.

True or False

1. _________ Saltwater fish get rid of excess water through osmosis.

2. _________ A fish uses its lateral line to help regulate it's buoyancy.

3. _________ All birds give birth to live young.

4. _________ Birds use their feathers to fly, to keep warm and dry, and for display.

5. _________ The name amphibians comes from the Latin words "aqua" and "vida," meaning "water life."

6. _________ Amphibians can avoid predators by using camouflage or by using their poison glands.

7. _________ Reptiles must move from warm places to cool places to control their body temperature.

8. _________ Some reptiles lay eggs; some reptiles give birth to live young.

9. _________ Monotremes are the only mammals to lay eggs.

10. _________ Mammals are the quickest of the chordates to reproduce.

Short Answer

1. How does a fish breathe in the water?

2. What are three characteristics that enable birds to be so efficient at flight?

3. Briefly describe the amphibian life cycle (or draw the life cycle of a frog).

4. What are the four main groups of reptiles?

5. Explain how the three groups of mammals differ in the way that they reproduce.

6. What are the five basic classes of the Phyla Chordata? Give the basic characteristic of each.

Planet Earth Unit Test

Vocabulary Matching

1. Biome ___
2. Cartographer ____
3. Coast ___
4. Delta ___
5. Deposition ___
6. Erratics ___
7. Estuary ___
8. Glacier ___
9. Greenhouse gas ____
10. Lines of Latitude ____
11. Lines of Longitude ____
12. Mantle ____
13. Moraine ___
14. Natural cycle ____
15. Oceanic ridge ___
16. Oceanic trench ___
17. Source ___
18. Tides ___

A. Lines that run from the North to the South pole, dividing the globe into segments; the lines all meet at the poles.

B. The exchanging of essential elements, such as nitrogen, carbon, and oxygen.

C. The mostly solid part of the Earth that lies between its crust and its core.

D. A fan-shaped system of streams that is created when a river splits into many smaller branches before it enters the sea.

E. Any gas that traps heat from the Sun, such as carbon dioxide.

F. A person who makes maps.

G. The dropping or leaving behind of rock and other debris by a glacier.

H. Lines that run parallel around the globe, dividing the globe into flat slices; the lines never meet.

I. A mass of ice that gathers at the top of a land mass and slowly flows downhill.

J. The beginning of a river.

K. A wide channel that forms where a river joins the sea.

L. A region of the Earth that contains unique plants and animals and is characterized by a distinct climate.

M. Rock, clay, sand, and other debris left by a glacier at the valley floor.

N. A deep trench in the seabed formed when one place of the Earth's crust moves under another.

O. The stretch of land that meets the sea.

P. The daily movement of the sea up and down the shore; they are caused by the gravitational pull of the moon.

Q. A raised ridge on the seabed caused by the movement of the Earth's crustal plates.

R. Large boulders that have been deposited away from their source by a glacier.

True or False

1.______________________ The Earth is magnetic due to the molten iron in its core.

2.______________________ The plates move because they float on top of the mantle.

3.______________________ Satellites do not help cartographers produce more accurate maps of the Earth.

4.______________________ Cartographers use projections to make flat maps of the Earth.

5.______________________ A river begins at the sea or ocean and flows inland to the mountains.

6.______________________ As a river slows, it deposits the heavier material it has picked up first.

7.______________________ Only one quarter of the Earth's surface is covered with salt water.

8.______________________ The gravitational pull of the Moon does not affect the oceans on Earth.

9.______________________ Glaciers can only be found on mountains.

10. ______________________ The presence of a deep, rounded, U-shaped valley and moraine hills let you know that a glacier was once present in the valley.

11. ______________________ In the carbon cycle, plants take in carbon dioxide and animals produce it.

12. ______________________ Pollution has no effect on the natural cycles.

13. ______________________ The boundaries of the different biomes are determined by variations in climate.

14. ______________________ A community is a group of plants and animals that live together in mutually beneficial relationships.

Short Answer

1. What are the 3 main layers of the Earth? (Be sure to include a detail or two about each.)

2. Label the globe below with the Arctic Circle, Prime Meridian, Equator, Tropic of Cancer & Tropic of Capricorn, North & South Pole.

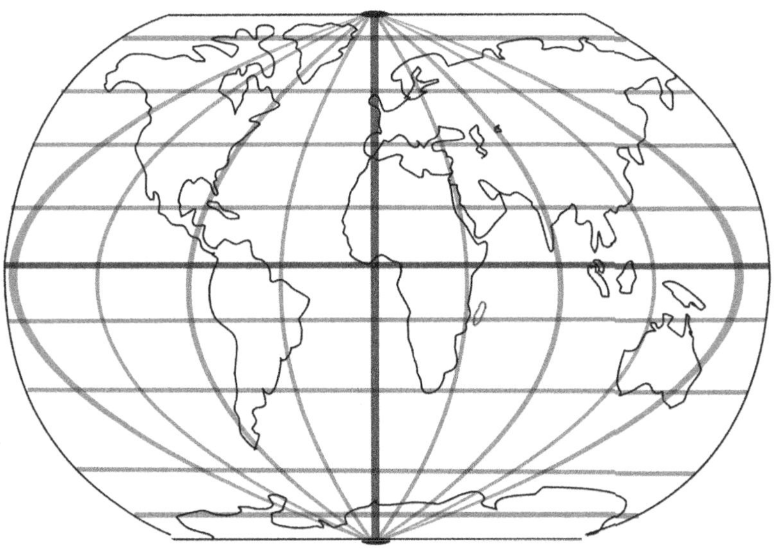

3. Describe the course of a river.

4. How do currents work in the ocean?

5. How do glaciers form?

6. Briefly describe the water cycle.

7. What is the difference between a habitat and a biome?

Made in the USA
Monee, IL
14 September 2026

62974519R00090